Maritime Strategy and National Security
in Japan and Britain

'Standing "shoulder to shoulder": General Sir Ian Hamilton with General Count Kuroki Tamemoto at the Battle of Shaho, 1904'.
Published with the permission of the Trustees of the Liddell Hart Centre for Military Archives, King's College London.

Maritime Strategy and National Security in Japan and Britain

From the First Alliance to Post-9/11

Edited by

Alessio Patalano
King's College, London, UK

GLOBAL
ORIENTAL

LEIDEN • BOSTON
2012

The editor wishes to acknowledge the Great-Britain Sasakawa Foundation, the Daiwa Anglo-Japanese Foundation, and the Japan Foundation for their generous financial contribution towards the realisation of this volume.

Cover illustration (clockwise from top left): RAF Harrier embarked on HMS Ark Royal. © UK MOD Crown Copyright [2010]; Japanese battleship *Hatsuse*, built on the Tyne at the Elswick ship-building yard, passing through the Swing Bridge, Newcastle-upon-Tyne, 1901. Photo: Courtesy Vickers Defence Systems; HMS Ark Royal visiting HM Naval Base Clyde in Scotland. © UK MOD Crown Copyright [2010]; Japan Maritime Self-Defence Force SH-60 taking off for a night flight. Courtesy of the Japan Maritime Self-Defence Force (2009).

Library of Congress Cataloging-in-Publication Data

Maritime strategy and national security in Japan and Britain : from the first alliance to post-9/11 / edited by Alession Patalano.
   p. cm.
 Includes index.
 ISBN 978-1-906876-27-2
 ISBN 1-906876-27-4
   1. Sea power–Japan–History. 2. Sea power–Great Britain–History. 3. Pacific Ocean–Strategic aspects. 4. Japan–Relations–Great Britain–History. 5. Great Britain–Relations–Japan–History.

 VA653.M33 2012
 359'.030941--dc23
                                                                        2012010791

ISBN 978 19 06 87627 2 (hardback)
ISBN 978 90 04 21262 6 (e-book)

MIX
Paper from
responsible sources
FSC® C109576
www.fsc.org

Printed by Printforce, the Netherlands

CONTENTS

PART THREE

MARITIME STRATEGY
IN AN INTERDEPENDENT WORLD

LIST OF MAPS, FIGURES AND TABLES

ACKNOWLEDGMENTS

In 2008, Japan and the UK celebrated 150 years of diplomatic relations. To celebrate the military dimension of these ties, the Department of War Studies, King's College London, and the Embassy of Japan in London co-organized a one-day conference titled 'Seizing the trident, drawing the sword: Anglo-Japanese military relations from the alliance to an inter-dependent world'. The conference was held on 27 May 2009 and it was well attended. The topic, focusing on matters of strategy and military affairs, was relatively 'unusual' for an academic conference on Japan and the UK and therefore, attracted a very diverse audience. It included many undergraduate and postgraduate students from the East Asian security and naval history programmes taught at the Department of War Studies, academics from a variety of universities, journalists, and regional experts from the world of think tanks as well as from military and diplomatic circles.

The date of the event itself was chosen with the intent to emphasize the depth and historical nature of UK-Japan military relations. On the same day, one hundred and four year earlier, the flagship of the Imperial Japanese Navy, the British-built battleship *Mikasa*, had led the Emperor's fleet to a staggering victory against a major European power, Russia. This one single battle became an iconic event, marking the country's successful completion of half a century of continued efforts towards modernization. Military ties between Japan and the UK, especially in their naval component, had played a crucial role in the events that led to this defining phase of Japanese modern history. It is perhaps symptomatic that, as the Japanese warships sailed into battle, the fleet's Commander-in-Chief, Admiral Tōgō Heihachirō, sought to boost the morale of his force in a clear Nelsonian fashion. He ordered to hoist the 'Z' flag signal from the yardarm of *Mikasa*. The message read: 'The fate of the Empire rests on this battle. Let every man do his utmost.' This reference to the flag signal used by Admiral Nelson at the battle of Trafalgar, was not casual. Admiral Tōgō, like many of his officers and of his predecessors, had learned the core elements of the business of modern naval warfare in the UK, and British officers and technology had helped form the backbone of the Japanese navy.

The idea for this conference was born during a three month-long research stay in Japan in the summer of 2007. I was in Tokyo conducting fieldwork activity for my doctoral thesis and in many of the interviews I conducted, I was struck by the way senior Japanese scholars and military officers often spoke of the British military experience. There was great admiration for it, and a sense of curiosity for a country that they considered having many similarities with Japan. After all, both Japan and the UK were and are, 'two island nations' – many commented to me. Militarily, however, this notion seemed to be applicable only to the pre-war period, and the navy-to-navy relationship was one of the first components to be mentioned.

What this notion exactly meant in the contemporary world was a different question. The meaning was never too clear, and people had very different views of what this notion entailed, if anything at all. In my conversations with Japanese and British friends and colleagues, it was often the case that similarities in military affairs were either dismissed (both sides pointing out the constitutional limitations that made the Japanese and British experiences very different), seemed to be limited now to the struggle to cope with reduced military capabilities, in comparison to the pre-war era, and a shared strategy of alignment with the United States. As a research student of Japanese post-war naval strategy and defence policy, I did not necessarily share this view and it seemed to me that this wider lack of agreement on the subject deserved more in-depth examination.

A number of individuals and institutions, both in Japan and the UK, supported me in the quest to implement the original idea and make the conference happen. At King's College, my two supervisors, Professor Andrew Lambert and the late Professor Saki Dockrill, discussed with me the structure of the conference, the themes to explore, and constantly provided me with suggestions, feedback and an overall sense of direction. Professor Lambert had been the person to introduce me to the foundations of British naval history and strategy, and Saki-san shared with me her understanding of Japan. Together they helped me make the conference an event that could bridge the gap between Anglo-Japanese studies and contemporary strategic affairs. Professor Mervyn Frost, Head of the Department of War Studies, made sure that a young research student such as myself could enjoy all the necessary institutional support from King's College, whilst Ms Jayne Peake, the Department's event manager, secured a most charming venue for the conference at the Maughan Library in Chancery Lane.

In Japan, I am particularly grateful of the constant assistance and encouragement I received from Professor Haruo Tohmatsu, the National Defence Academy, and Professor Chiyuki Aoi, Aoyama Gakuin University. Tohmatsu-sensei was a star in pointing me towards the right founding bodies and in coaching me to present the topic in an appropriate fashion. Aoi-sensei, with her interest in the British Army's doctrinal evolution, was an endless source of intellectual stimulus, introducing me to new perspectives on how the British experience can be of help to a fast-changing Japanese Ground Self-Defence Force. Dr Ken Kotani and Tomoyuki Ishizu, both at the National Institute for Defence Studies and both deeply knowledgeable of British and Japanese military history, offered insights and suggestions on possible themes to examine. In the Japanese armed forces, a number of mid-career, senior and retired officers were kind enough to share their views with me. In particular, Vice Admiral Yōji Kōda, JMSDF (Ret.), and Lieutenant General Noboru Yamaguchi, JGSDF (Ret.) were very patient and approachable, sparing long hours of conversations to introduce me to the recent, and less recent, military history of Japan. I am equally grateful to late Captain Tsubasa Kannō, JMSDF, who gave great value to the underpinnings of British and Japanese strategy and decided to come to the Royal College of Defence Studies to study them.

A number of other institutions supported the entire project with great enthusiasm and encouragement. The Japanese Embassy in London was never too far away and kindly offered to co-organize the event, and offer financial support to it. The event was registered with the 'Japan-UK 150' initiative and was awarded additional financial contributions from the Daiwa Anglo-Japanese Foundation, the Great Britain Sasakawa Foundation and the Japan Foundation. The individuals I interacted with at all these institutions showed great professionalism, patience and faith in a young – untested-scholar. For this I am most grateful.

Part of the funding was aimed at the publication of the event's proceedings. This volume represents the culminating phase of the original event. In the preparation of the final manuscript, I profited from my conversation with Mr Paul Norbury at Brill/Global Oriental to sharpen some of the ideas in the volume. Above all, I am most grateful for his unfettered commitment to the project since we first talked about it at a coffee table at the IOD late in the Spring 2009. Paul's assistant editor, Ms Nozomi Goto, was always extremely efficient and kind as I sent her the different components of the manuscript. An edited volume is always a joint-effort. This project represents the summation of the intellectual contributions of leading scholars in the field of British and Japanese military history and strategy and uniformed practitioners from the UK and Japan with extensive experience in the areas of strategic planning and defence policy-making. It was an honour and a pleasure to work with them and any misrepresentation or shortcoming related to it rests upon my shoulders.

The summer that followed the conference was one of unexpected dramatic events. Two of the individuals who had inspired me to pursue this project, Professor Saki Dockrill and Captain Tsubasa Kannō, were prematurely taken away from the love of friends and colleagues. I lost two mentors and friends. Throughout their lives, both had contributed greatly to keep Japan and the UK in each other's intellectual and military horizons. In my conversations with them, a genuine sense of respect for the common native land, Japan, and a country they both admired, the UK, permeated their words and thoughts. Both passed away long before they could fully express the contribution they had to give to the understanding and promotion of Japan-UK relations. This volume is dedicated to their memory.

Alessio Patalano
London, 15 June 2011

ABBREVIATIONS AND ACRONYMS

ACSA	Acquisition and Cross-Servicing Agreement between Japan and the United States
AQ	Al Qaeda
ASW	Anti-Submarine Warfare
ATSML	Anti-Terrorism Special Measures Law
AWACS	Airborne Warning and Control System
BA	British Army
C4I	Command, Control, Communications, Computers and Information
CENTCOM	Central Command, United States
CinC	Commander in Chief
CIWS	Close-In Weapons System
CLB	Cabinet Legislative Bureau, Japan
CMS	Chief of Maritime Staff, JMSDF
CONMAROPS	Concept of Maritime Operations, NATO
COS	Chiefs of Staff, British Armed Forces
CRF	Central Readiness Force, JSDF
CTF-150	Combined Task Force-150
DDH	Helicopter Carrying Destroyer
DPJ	Democratic Party of Japan
EASR	East Asia Strategy Report
EEZ	Exclusive Economic Zone
FECB	Far Eastern Combined Bureau
FY	Financial Year (1 April-31 March)
GDP	Gross Domestic Product
GHQ	General Headquarters
GLCMs	Ground-Launched Cruise Missiles
GSO	Ground Staff Office, JGSDF
HADR	Humanitarian Assistance and Disaster Response
IJA	Imperial Japanese Army
IJN	Imperial Japanese Navy
IPC	International Peace Cooperation Law
IPCATU	Peace Cooperation Activities Training Unit
JAM	Jaysh Al Mahdi
JASDF	Japan Air Self-Defence Force

JCG	Japan Coast Guard
JDA	Japan Defence Agency
JGSDF	Japan Ground Self-Defence Force
JIC	Joint Intelligence Committee
JMoD	Japan Ministry of Defence
JMSDF	Japan Maritime Self-Defence Force
JSDF	Japan Self-Defence Forces
JSO	Joint Staff Office, JSDF
LST	Tank Landing Ship
MCM	Mine Counter-Measures
MEW	Ministry of Economic Warfare
MINUSTAH	United Nations Stabilization Mission in Haiti
MOF	Ministry of Finance
MOFA	Ministry of Foreign Affairs
MOOTW	Military Operations Other Than War
MRBMs	Medium-Range Ballistic Missiles
MSO	Maritime Staff Office, JMSDF
MSR	Main Supply Routes
MTDP	Mid-Term Defence Programme
NATO	North Atlantic Treaty Organization
NDA	National Defence Academy
NDPG	National Defence Programme Guideline
NDPO	National Defence Programme Outline
NEP	National Evacuation Operation, UK
NPR	National Police Reserve
NSF	National Safety Force
ODA	Official Development Assistance
OEF	Operation Enduring Freedom
OEF-MIO	Operation Enduring Freedom, Maritime Interdiction Operation
ONUMOZ	United Nations Operations in Mozambique
PAC-3	Patriot Advance Capability-3
PKO	Peace-keeping Operations
PLAN	People's Liberation Army Navy
POWs	Prisoners of War
RAF	Royal Air Force
RFA	Royal Fleet Auxiliary
RN	Royal Navy
SALT	Strategic Arms Limitation Talks
SCC	Japan-US Security Consultative Committee

SDR	Strategic Defence Review, UK
SLBMs	Submarine Launched Ballistic Missiles
SLOCs	Sea Lanes of Communication
SOP	Standard Operational Procedures
SSBN	Ballistic missile submarine
SSM	Surface-to-Surface Missile
SSN	Nuclear-powered Attack Submarine
SSR	Security Sector Reform
SVTOL	Short and Vertical Take-Off and Landing
TICAD	Tokyo International Conference on African Development
UKNA	United Kingdom National Archives
UNCLOS	United Nations Convention for the Law of the Sea
UNCMA	United Nations Civil-Military Affairs
UNDOF	United Nations Disengagement Observer Force
UNDPO	UN Department of Peacekeeping Operations
UNMIN	United Nations Mission in Nepal
UNMIS	United Nations Mission in Sudan
UNPROFOR	United Nations Protection Force
UNSAS	UN Standby Arrangement System
UNTAC	United Nations Transitional Authority in Cambodia
UNTAET	United Nations Transitional Administration in East Timor
USN	United States Navy
WMD	Weapons of Mass Destruction

CONTRIBUTORS

Professor Chiyuki Aoi is Associate Professor of International Politics at Aoyama Gakuin University, Japan. She is also lecturer at the Joint Staff College and the National Institute for Defence Studies, Japan Ministry of Defence. Her research focuses on stability operations, Japanese peace cooperation and the legitimacy of the use of armed forces in post-Cold War stability missions. Her works include *Unintended Consequences of Peacekeeping Operations,* co-authored with Cedric de Coning and Ramesh Thakur (2007).

Dr Guibourg Delamotte is lecturer in International relations and Japanese politics at the National Institute for Oriental Languages and Civilizations (INALCO), France. A graduate of the universities of Paris and of Oxford where she read Law, she completed her PhD at the Paris-based School of Social Sciences (EHESS) in 2007. Her dissertation on Japan's defence policy received the Shibusawa-Claudel award in 2008.

Professor John Ferris is former Head of the History Department at The University of Calgary. He specializes in military and diplomatic history as well as in intelligence. He has written widely about British diplomacy and strategy between the 1870s and 1945, and published on topics such as sea power between the World Wars, American intelligence and Western perceptions of Japanese military power.

Dr Douglas Ford joined the University of Salford in 2004 as a lecturer in military and international history. He holds MA and PhD degrees from the London School of Economics, and has published over a dozen scholarly works on British and American intelligence on the Imperial Japanese armed forces during the Pacific War. His latest book, *The Elusive Enemy: U.S. Naval Intelligence and the Imperial Japanese Fleet* was published in 2011 by Naval Institute Press.

Professor Eric Grove is Professor of Naval History and Director of the Centre for International Security and War Studies at the University of Salford. His areas of expertise encompass maritime and naval history and strategy. He is a prolific author, his major works being *Vanguard to*

Trident, British Naval Policy Since 1945 (1987), *The Future of Sea Power* (1990), and *The Royal Navy Since 1815* (2005).

Commodore Steven Jermy, Royal Navy (Ret.), first saw active service in the Falklands in Op CORPORATE, flying in Sea King helicopters. His career has seen a mix of active duties and staff appointments. Staff appointments included Assistant Director (Strategy) on the MoD Naval Staffs, Deputy Director for Policy Planning on the MoD Central Staffs, and Principal Staff Officer to Chief of Defence Staff. In addition to Op CORPORATE, he has been deployed operationally to Op KINGOWER (Balkans) and in 2007, Op HERRICK as the Strategy Director in the British Embassy in Kabul.

Dr Alessio Patalano is lecturer in war studies at the Department of War Studies, King's College London, and specializes in East Asian security and Japanese naval history and strategy. Since 2006, he is visiting lecturer at the Italian Naval War College (ISMM), Venice. In Japan, Dr Patalano has been a visiting scholar at Aoyama Gakuin University and at the National Graduate Institute for Policy Studies (GRIPS), both in Tokyo, and currently is adjunt fellow at the Institute of Contemporary Asian Studies, Temple University Japan. Dr Patalano's publications appeared in academic journals in English, Japanese and Italian language.

Haruo Tohmatsu is Professor of diplomatic and war history at the National Defense Academy of Japan. He holds degrees from the University of Tsukuba and Waseda University and a D. Phil from the University of Oxford. His research interests encompass the history of Anglo-Japanese relations, the Russo-Japanese War, the Manchurian Incident, Japan's Micronesian mandates, the Sino-Japanese War of 1937-1945. As a member of the editorial board of the Military History Society of Japan (MHSJ) he recently edited a book entitled *The Sino-Japanese War of 1937-45 Revisited* (Kinseisha, 2008). In English, he co-authored numerous books including *A Gathering Darkness: The Coming of War to the Far East and the Pacific, 1921-1942* (Rowman & Littlefield, 2004), and *Imperialism on Trial: The International Oversight of Colonial Rule* (Lexington Books, 2006).

Lieutenant General Noboru Yamaguchi, JGSDF (Ret.), is professor of military history and strategy at the National Defence Academy of Japan. General Yamaguchi served as Senior Defence Attaché at the Japanese

Embassy in the United States, as Director for Research of the Ground Research and Development Command (GRDC), and as Vice President of the National Institute for Defence Studies. From 2006, he was the Commanding General of the GSDF Research and Development Command until he retired from active duty in December 2008.

Vice Admiral Yōgi Kōda retired from the JMSDF in August 2008 as Commander in Chief, Self-Defence Fleet. His prior assignments included tours of duty commanding surface units as well as posts at the Maritime Staff Office. Admiral Koda served as Director of the JMSDF's Planning and Policy Division and was directly involved in the first deployment of Japanese destroyers in the Indian Ocean. He has published several articles in English on Japanese naval history and strategy in the *Naval War College Review*.

CONVENTIONS

This volume adopts the revised Hepburn Romanization system. Macrons are used to indicate long vowels in Japanese, except for place names (e.g. Tokyo, Kyoto, Osaka) in common use in English language publications.

Japanese names are given with family names preceding first names. In bibliographical references, names of Japanese authors are given according to Western practice. Names of Japanese authors in English language publications are reported according to the original text.

In the text, reference is made to the conventional periodization of Japanese modern history, the system of 'era names',[1] which is used in Japan instead of the Gregorian calendar. *Nengō* derive their names from the ruling Emperor. Since 1868, Japanese history has been distinguished by four eras:

- Meiji (1868-1912);
- Taishō (1912-1926);
- Shōwa (1926-1989);
- Heisei (1989-Present).

---

[1] (Nengō – 年号).

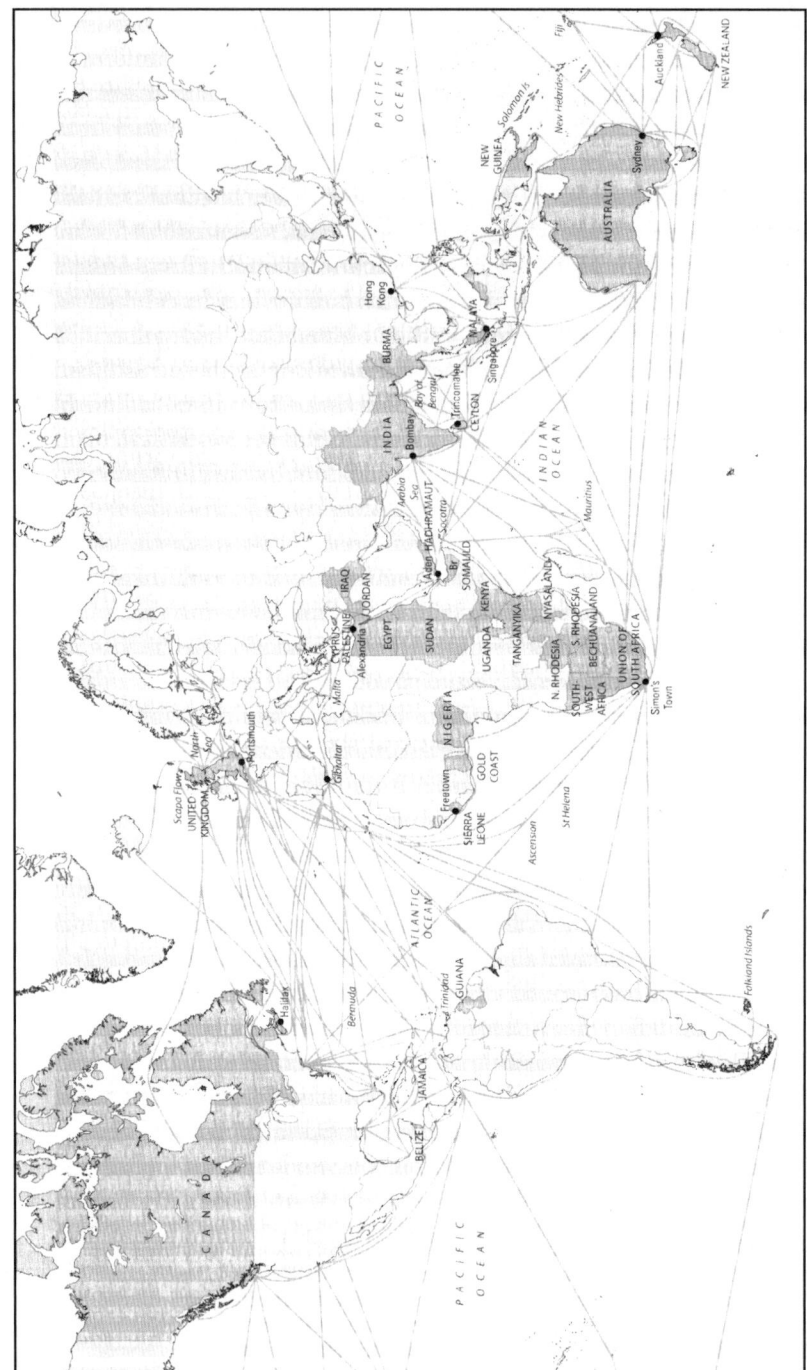

I. The British Empire, 1939

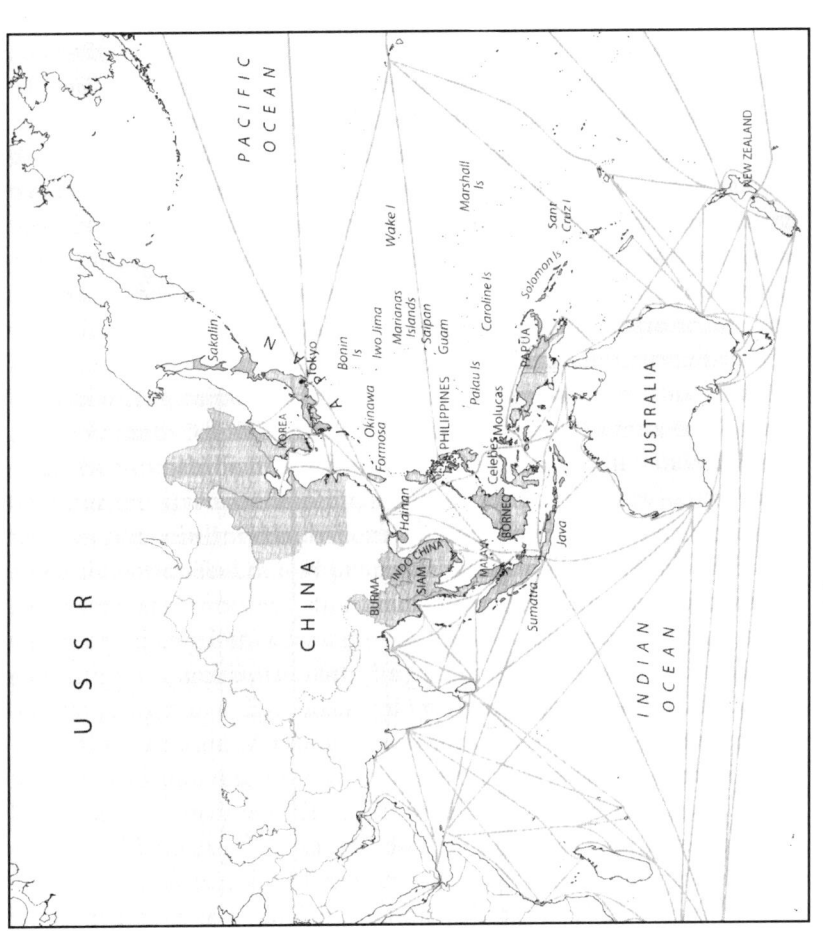

II. The Japanese Empire, 1942

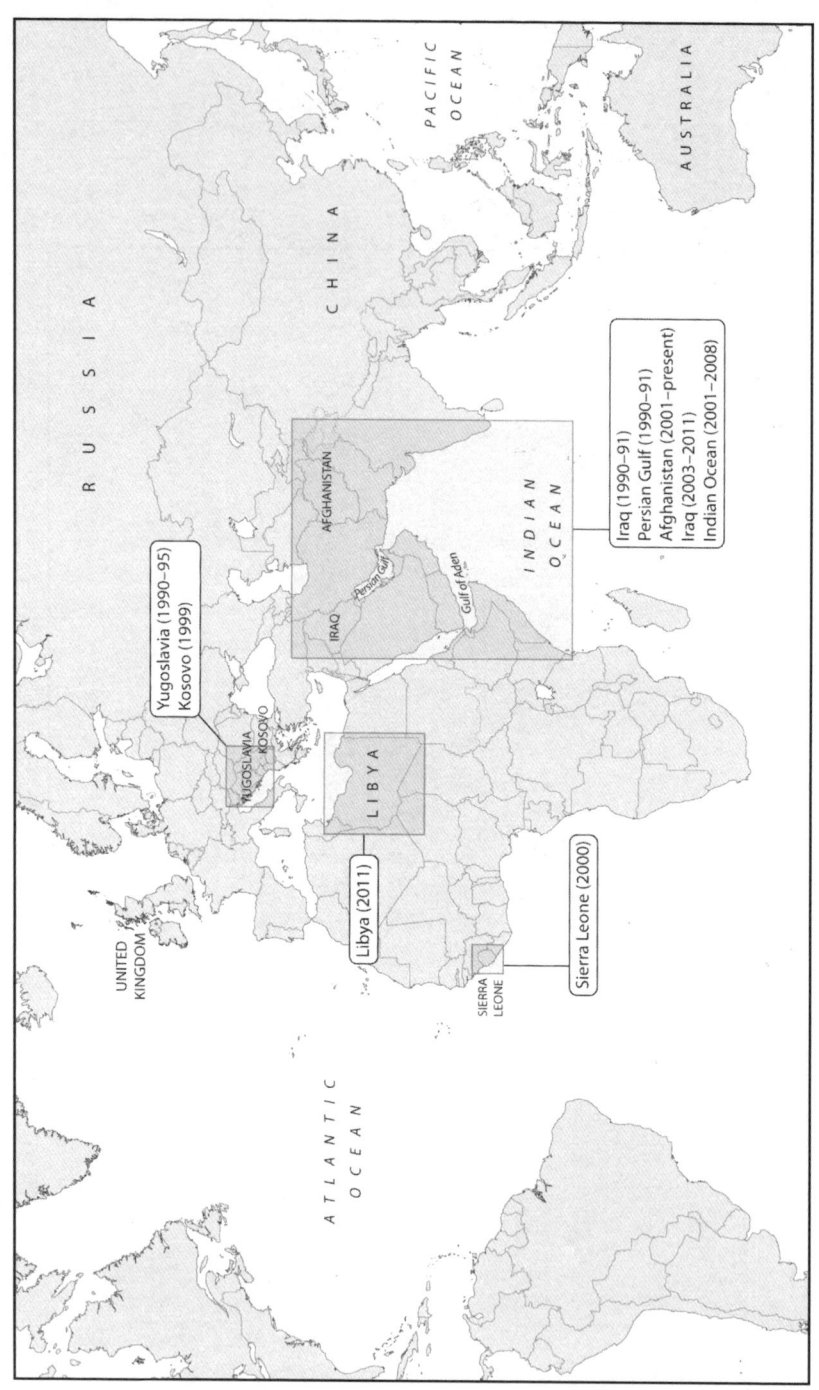

III. Main British military operations, 1990s and 2000s

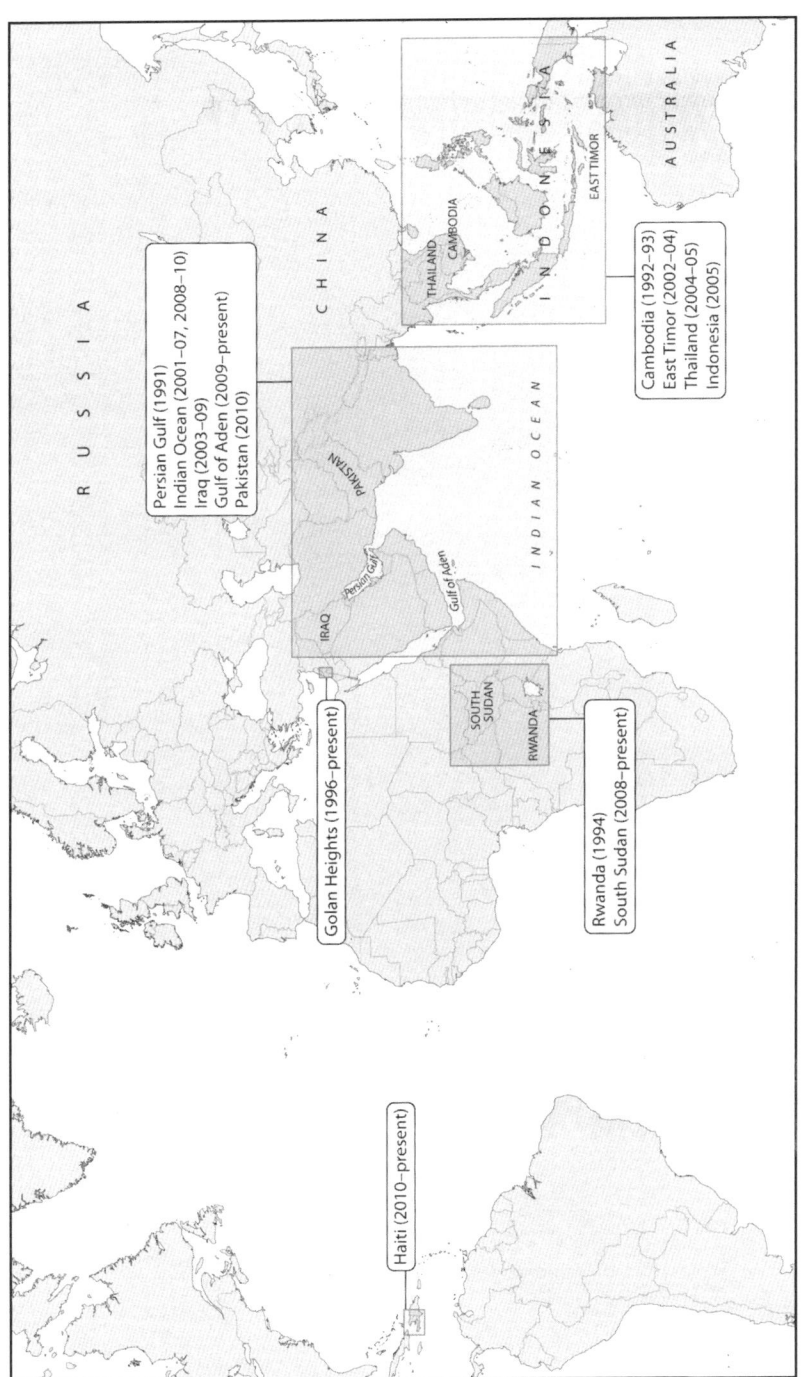

IV. Main Japanese military operations, 1990s and 2000s

MARITIME STRATEGY AND NATIONAL SECURITY IN JAPAN AND BRITAIN

Alessio Patalano

A 1969 study of the first decades of interaction between Japan and the UK pointed out that when formal diplomatic relations were established, the 'two island empires' had very little in common. In the author's view, the differences were many, substantial and could be summarized as follow:

> Britain, a world-wide empire dedicated to free trade; Japan, a group of four islands, almost isolated from the world for more than two centuries by a self-imposed seclusion policy. Britain, a constitutional monarchy run by a parliamentary system; Japan, a decadent military dictatorship. Britain, a rich strong manufacturing nation; Japan, a traditionally feudal society undermined by a developing money economy. Britain, a strong-hold of Protestant evangelism; Japan, since 1616 the declared persecutor of Christians. Britain, the advocate of an informed public opinion; Japan, a nation without a press, her masses generally denied political knowledge.[1]

This tinted vision of their differences notwithstanding, in two respects the author considered Japan and the UK very similar. 'Each people had a rich but vastly different cultural heritage, intense national consciousness, and an assurance of superiority. Each had long felt the security of insularity while profiting by the cultural influences of a neighbouring continent'.[2] At a point in time when Japanese and British social, economic and political structures were as opposite as night is to day, geography still drew the two countries close together. Their similar 'insularity' meant in fact that the sea had had a crucial strategic role in national security. Whether as a staging platform for imperial trade

---

[1] Grace Fox, *Britain and Japan, 1858-1883* (Oxford: Clarendon Press, 1969), 3.
[2] Ibid.

and expansion, or as a 'stopping force' against foreign invasions,[3] the way Japanese and British authorities approached the use of the sea had a direct impact on the policies they developed to pursue the defence of the national realm.

What about today? One hundred and fifty-three years after diplomatic relations were established in 1858, is geography still relevant in matters of strategy and national security in Japan and the UK? Geography hardly ever changes and therefore, instinctively, one would feel tempted to answer 'yes, it does matter'. Yet, across more than a century, many other external and domestic factors affecting national security policy, including the international landscape, national political orientations, economic outlook, technological know-how and industrial output have changed. How has the relationship between the strategic use of the sea and national security evolved since the second half of the nineteenth century? Does 'insularity' maintain a meaning that is distinctive to the strategies underpinning Japanese and British security? Does sea power still matter to Japanese and British political and military calculations to develop sound defence policies? How should this relationship inform military relations between Japan and the UK?

This book addresses the above questions, exploring the role of sea power in shaping Japanese and British national defence policy across the past century and a half. In this book, sea power is referred to in its 'historical/cultural' dimension. As a notion, 'sea power' lacks a universally agreed definition. There are, however, two main ways to understand it. The first falls within the intellectual tradition established by American strategist and naval educator Captain Alfred T. Mahan, US Navy (USN). It focuses on the 'strategic' and military dimensions of sea

---

[3] American realist scholar John Mearsheimer argued that the sea possesses an inherent 'stopping power', imposing severe limitations to state actors in their ability to wage war across extended water masses. This notion has been criticized in more recent scholarly work for it underestimated the use of the sea, throughout history, to project power away from national shores. In this context, the key point is that in pre-modern times, the geographic distance existing between Japan and the neighbouring Asian landmass – greater than that existing between the UK and the European continent – made an invasion of the archipelago a rather expensive and logistically complex affair. Until the nineteenth century, only the Mongol expeditions of 1274 and 1281 threatened the independence of the Japanese archipelago. In pre-modern times, the political and financial costs to develop and maintain the technological know-how to project forces and sustain them across the seas represented a key factor in the 'stopping power' of the sea. John J. Mearsheimer, *The Tragedy of Great Power Politics* (New York: Norton, 2001). For a critique of Mearsheimer's notion, Jack S. Levy and William R. Thompson, 'Balancing on Land and at Sea: Do States Ally against the Leading Global Power?', *International Security*, Vol. 35, 2010:1, 7-43.

power, investigating the function of navies and naval warfare as a tool of national strength.[4] The second refers to a 'historical/cultural' construction of sea power, first developed by British historians and imperial strategists like George Grote, John Robert Seeley and Sir Julian S. Corbett. In it, the core of the analysis does not focus on the means used to operate at and from the sea. Instead, the emphasis is on the nation's dependence upon the use of the sea for its economic survival. In this view of sea power, naval forces are a function of a 'maritime strategy' that seeks to harmonize strategic geography and military power (both land and naval) for the pursuit of economic benefit.[5] In it, the size and shape of naval forces is functional to the country's ability to defend territorial integrity and commercial maritime enterprises, putting a premium on deterrence to favour trade and commerce.[6] In the words of Geoffrey Till, this second approach is about 'the sea-based capacity of states to shape events both at sea and on land', it is about the effects and consequences produced by the use of the sea to enhance national security.[7]

Based on this understanding of sea power, the contributors to this volume were not asked to engage in an analysis of the configuration and reconfiguration of Japanese and British naval forces, the armed element of sea power. They were asked to engage with the wider political and doctrinal processes underpinning shifts in Japanese and British appreciation of their 'insularity' and the ways in which these informed the evolution of national strategy. From different perspectives, the authors debated the evolution of roles and missions of military forces in relation to the wider foreign and security goals they were meant to serve, in war as well as in peacetime. In this respect, the chapters in this book look at the ways in, and the extent to, which a 'maritime strategy' has

---

[4] Mahan's core ideas are set forth in his best known work, *The Influence of Sea Power upon History, 1660-1783* (Boston, MA: Little Brown, 1890). For a comprehensive introduction to Mahan's works, cf. John T. Sumida, *Inventing Grand Strategy and Teaching Command: The Classic Works of Alfred Thayer Mahan* (Baltimore, MD: Johns Hopkins University Press, 1997).

[5] For definitional clarifications on the adjectives 'naval' and 'maritime', see Ian Speller, 'Naval Warfare', in David Jordan, James D. Kiras, David J. Londsdale, Ian Speller, Christopher Tuck and C. Dale Walton, *Understanding Modern Warfare* (Cambridge: Cambridge University Press, 2008), 125.

[6] Andrew Lambert, 'Sea Power', in George Kassimeris and John Buckley (eds.), *The Ashgate Research Companion to Modern Warfare* (Farnham, Surrey: Ashagte, 2010), 73-87.

[7] Geoffrey Till, 'Introduction: Sea power and the Rise and Fall of Empires', in G. Till and Patrick C. Bratton (eds.), *Sea Power and the Asia-Pacific: The Triumph of Neptune?* (Abingdon, OX: Routledge, 2011), 2.

been, and continues to be, instrumental to Japanese and British security and international profile.

Maritime Strategy and the Notion of Japan
as the 'Britain of the Far East'

Writing about maritime strategy and security policy in Japan and the UK requires an engagement with the notion of 'Japan as the Britain of the Far East'. It is unclear where this idea was first used. It is known that by the opening decade of the twentieth century, the analogy was regularly accepted in Britain to teach pupils about Japan and its geography, and referred to by prominent public figures to explain its modernization and international ambitions.[8] In Japan, the comparison with Britain was frequent, though not exclusive, and it was not uncommon for Japanese diplomats and political figures to use it to attract the attention of British counterparts on the archipelago's strategic significance in the fast evolving East Asian regional context.[9] Scholars suggested that in Japan the original use of this analogy appeared as a pragmatic choice. In the second half of the nineteenth century, British behaviour in international affairs offered Japanese elites a valuable template regarding the options open to an island nation. 'Even before the treaties and during their early years Japanese officials and scholars had chosen Britain as their nation's model.' Reformers of the Bakumatsu and Meiji eras '(r)ecognizing the similarity of Japan's geographical relation to Asia to that of Britain's to Europe, and wishing to emulate British achievements, they set about making Japan the Britain of the Far East'.[10]

At the heart of the analogy rested a formula for the harmonization of insular geography, economic outlook and national security. The interaction of these factors informed defence policy as well as the build up and use of military power. Lord George Curzon, who travelled to Japan and observed the country's transformation at the end of the nineteenth

---

[8] Charles A. Fisher, 'The Britain of the East? A Study in Geography of Imitation', *Modern Asian Studies*, Vol. 2, 1968:4, 343-344. In 1894, one of the first publications to popularize the analogy in the English language was George N. Curzon, *Problems of the Far East: Japan, Korea, China* (London: Longmans, 1894).

[9] Fisher, 'The Britain of the East?', 348-356. On the use of the analogy for political purposes, cf. Manjiro Inagaki, *Japan and the Pacific and a Japanese View of the Eastern Question* (London: Fisher Unwin, 1890).

[10] Fox, *Britain and Japan*, 532.

century, summarized what it meant for Japan to aspire to be the 'Britain of the Far East' as follows:

> Placed at a maritime coign of vantage upon the flank of Asia, precisely analogous to that occupied by Great Britain on the flank of Europe, exercising a powerful influence over the adjoining continent, but not necessarily involved in its responsibilities, she sets before herself the supreme ambition of becoming, on a smaller scale, the Britain of the Far East. By means of an army strong enough to defend our shores, and of a navy sufficiently powerful to sweep the seas, she sees that England has retained that unique and commanding position in the West which was won for us by the industry and force of character of our people, by the mineral wealth of these islands, by the stability of our Government, and by the colonising genius of our sons. By similar methods Japan hopes to arrive at a more modest edition of the same result in the East.[11]

As this passage makes clear, from a military point of view, being the 'Britain of the Far East' was not just about possessing a strong navy. It was about developing a defence posture centred on a finely balanced mix of land and naval forces (a mix that today would necessarily include air forces). Such a military balance was defined by the nature of the threat posed by the neighbouring continent and by the nature of its economic interests beyond national shores. In this context, the sea had a crucial role in national security. It acted as a geophysical defensive shield, as a platform for the exercise of deterrence and, if necessary, as an enabling network of far-reaching waterways available to project power to preserve vital economic and security interests.

After the signing of the Anglo-Japanese Alliance in 1902 and until 1907, Japan continued to be regarded as the mirror image of Britain in East Asia, especially since the bilateral partnership was in essence a naval alliance. For Britain, the Japanese navy was an essential partner to keep the Russian menace against the British Empire in Asia in check.[12] For Japan, Ian Nish observed, 'the alliance conferred status, recognition of her aspirations in the far east, and the co-operation of the largest naval power in the world'.[13] After 1907, the disappearance of Russia as

---

[11] Curzon, *Problems of the Far East*, 395-6. Lord Curzon's involvement with Japan is examined in Ian Nish, 'Lord Curzon (1859-1925) and Japan', in Hugh Cortazzi (ed.), *Britain and Japan: Biographical Portraits* (Vol. V, Folkstone, Kent: Global Oriental, 2005), 96-106.

[12] Keith Neilson, 'The Anglo-Japanese Alliance and British Strategic Foreign Policy, 1902-1914', in Phillips Payson O'Brien, *The Anglo-Japanese Alliance, 1902-1922* (London: Routledge, 2004), 51-52.

[13] Ian Nish, *The Anglo-Japanese Alliance: The Diplomacy of Two Island Empires, 1894-1907* (London: The Athlone Press, 1966), 373.

the main common threat changed the nature of the alliance and the use of the analogy slowly lost its original meaning. In the 1930s, Japanese national strategy took a decisive turn away from the 'island nation' model it was originally thought to follow.[14] The country's political and military elites embarked upon an aggressive continental strategy and on a more substantial involvement in China. In Britain, Japanese penetration on the continent contributed to change perceptions, and Japan became to be regarded as a threat to the British Empire.[15] Japan was no longer the 'Britain of the Far East'.

In the post-1945 period, the initial legal and political conditions in Japan prevented any sort of comparison with Britain. Similarly, in Britain, wartime grievances related to the treatment of prisoners of war (POWs), issues of reconciliation and economic competition, set the tone of Anglo-Japanese interactions in the initial post-war years.[16] In the 1960s, University of Kyoto Professor Kōsaka Masataka was the first academic to re-introduce into the Japanese public debate the idea of Japan as an 'island nation' (*shimaguni*). In his writings he argued that the country's miraculous economic recovery in the 1950s and expansion in the 1960s rested on the ability to use the sea for trade and commerce.[17] Kōsaka did not seek to develop his vision of a 'maritime country' (*kayō kokka*) into a military strategy. Its essence permeated the diplomatic and economic policies set forth by Prime Minister Yoshida Shigeru in the mid-1950s and pursued by subsequent Japanese Prime

---

[14] Fisher, 'The Britain of the East?', 366-368.

[15] See Anthony Best, 'The Road to Anglo-Japanese Confrontation, 1931-1941', in Ian Nish and Yoichi Kibata (eds.), *The History of Anglo-Japanese Relations, 1600-2000, Volume II: The Political-Diplomatic Dimension, 1931-2000* (London: Palgrave MacMillan, 2000), 26-50.

[16] Tanaka Akihiko, 'Anglo-Japanese Relations in the 1950s: Cooperation, Friction, and the Search for State Identity', 201-234; and Christopher Braddick, 'Distant Friends: Britain and Japan since 1958 – the Age of Globalization', 263-275, both in Nish and Kibata, *The History of Anglo-Japanese Relations, The Political-Diplomatic Dimension*; Peter Lowe, 'Britain and the Recovery of Japan Post-1945', in Philip Towle and Nobuko Margaret Kosuge, *Britain and Japan in the Twentieth Century: One Hundred Years of Trade and Prejudice* (London: I.B. Tauris, 2007), 97-106.

[17] See Kosaka Masataka, *Options for Japan's Foreign Policy*, Adelphi Paper 97 (London: International Institute for Strategic Studies 1973); the original idea was elaborated in Kōsaka Masataka, 'Kaiyō Kokka Nihon no Kōsō' [The Concept of Japan as Maritime State], *Chūō Kōron*, 1964:9, 48-80.

Ministers.[18] Yet, his work made a crucial point about the relevance of strategic geography to national security, linking sea power, in both its civilian and armed dimensions (the latter to safeguard the former), to economic survival. His views informed the 1970s defence policy review process that culminated in 1976 with the adoption of the National Defence Programme Outline (NDPO). In the 1980s, Japanese defence policy came to have a strong 'maritime flavour' and attempts to adapt national military power to neutralize the Soviet naval threat to Japanese shipping and sea lanes in the western Pacific became a mission of primary importance.[19]

By the end of the Cold War, Japanese efforts to protect the main sea lanes in the Sea of Japan and in the East China Sea, in the area between the archipelago and Taiwan to the south, and Guam to the southeast were functions of a strategy that compared well with that of the island nation model of British memory. Japanese military posture had limits but it was proportionate to the country's modest security goals. Above all, it benefited from the presence of a powerful regional ally, the United States. In this respect, as one informed observer recently noted, in the 1980s Japanese and American strategic priorities in regard to the Soviet threat were aligned. Japanese economic dependence upon the sea made a military posture with an appropriate naval and air balance a strategic imperative. In turn, Japanese sea power helped the United States 'to carry out its forward strategy in East Asia'.[20]

It is ironic that in the 1990s, the process of reaffirmation of the US-Japan alliance set the stage for the re-emergence of the 'Britain of the Far East' analogy, but with a very different meaning. In its new guise, the comparison with the UK was not based on strategic geography, but on the close military ties enjoyed by the UK with the United States. These

---

[18] Iokibe Makoto, 'Introduction: Japanese Diplomacy from Prewar to Postwar', in Iokibe Makoto (ed.), *The Diplomatic History of Postwar Japan* (Abingdon, OX: Routledge, 2011), 9. For a detailed assessment of Japanese diplomatic and economic initiatives in the 1960s and 1970s, cf. the chapters in the same volume by Tadokoro Masayuki, 'The Model of an Economic Power: Japanese Diplomacy in the 1960s', 81-107; and by Nakanishi Hiroshi, 'Overcoming the Crises: Japanese Diplomacy in the 1970s', 108-142.

[19] Alessio Patalano, 'Shielding the "Hot Gates": Submarine Warfare and Japanese Naval Strategy in the Cold War and Beyond (1976-2006)', *The Journal of Strategic Studies*, Vol. 31, 2008:6, 864-865, 867-868.

[20] George R. Packard, 'The United States – Japan Security Treaty at 50: Still a Grand Bargain?', *Foreign Affairs*, Vol. 89, 2010:2, 96. For an overall review of the transformation of alliance in the 1990s, Kent Calder, *Pacific Alliance: Reviving U.S.-Japan Relations* (New Haven, CT: Yale University Press, 2009).

were to be singled out as the core reason for Japan to look at the other end of the Eurasian continent for a model in security affairs.[21] In the more recent literature on Japanese security and defence policy, the tendency has been to focus on the use of this recent understanding of the analogy to set a standard against which policy shifts were measured.[22]

In this book, the original analogy is used as empirical reference – shifting the attention back to the question of the relationship between geography and strategy. What does this relationship entail in terms of Japanese and British national security policies? British scholar Colin Gray partly offered an answer to this question explaining the influence of geography on grand strategy in modern British history. As he pointed out, insularity did not 'mandate' Britain to maintain 'a superior naval fighting instrument'. It provided its political and military elites with a range of options. These included four types of strategies: a) involvement in continental military affairs to prevent the emergence of a major naval competitor; b) support of continental allies to limit the ability of potential competitors to threaten British interests at or from the sea; c) securing strong naval allies capable to protect the British Isles from invasion, and British commerce from harassment; d) maintaining a strong national fleet.[23]

According to Gray, British strategy over the past century and a half opted for a combination of the third and fourth options. Building up on these considerations, the chapters that follow seek to offer narratives that elucidate Japanese and British choices within their respective strategic contexts, evaluating policies, including the security alignment with the United States, against the need to adapt national agendas and capabilities to an insular geography.

National Security in Japan and Britain and Anglo-Japanese Relations

By focusing on parallel accounts of the evolution of Japanese and British strategy, this book is different from others dealing with Japan and the

---

[21] Alessio Patalano, 'Japan's Maritime Strategy: The Island Nation Model', *RUSI Journal*, Vol. 156, 2011:2, 82-83.

[22] Ibid., 83-84.

[23] Colin S. Gray, *Strategy and History: Essays on Theory and Practice* (Abingodn, OX: Routledge, 2006), 138. The chapter originally appeared as a journal article, titled: 'Geography and Grand Strategy', *Comparative Strategy*, Vol. 10, 1991, 311-329.

UK. Outside the alliance period – when bilateral relations occupied a significant place in their respective foreign and security policy – this book does not engage with an analysis of Anglo-Japanese bilateral relations. Contributors were not asked to directly comment on the state of bilateral military and security cooperation. Instead, the chapters are designed to propose a comparative study of Japanese and British national strategies as a way to establish whether areas of potential cooperation exist that are currently underdeveloped or unexplored. More importantly, by focusing on national narratives the book seeks to point out the strategic reasons for military and security cooperation to be considered.

For the alliance period, this book benefited greatly from the insights gained in the field of Anglo-Japanese relations by the groundbreaking work of scholars like Ian Nish, Philip Towle, Anthony Best, and of personalities like former Ambassador to Japan Sir Hugh Cortazzi. From a maritime perspective, the literature offered in-depth explanations of the reasons that drew the two countries close at the beginning of the twentieth century and what set them apart by the mid-1930s. Whilst it is outside the scope of this book to review this literature, it is important to stress that it contributed to detail Japanese and British evolving ambitions, military interactions, organizational influences, bilateral cooperation in the fields of intelligence and defence industry. This substantial body of knowledge was instrumental for the authors of this book to be able to focus on a debate about ideas and strategic views of the sea, rather than the uncovering of new evidence.

Recently, scholarship set out to expand the timeframe of the research agenda and started surveying the state of Anglo-Japanese relations after the war.[24] New studies assessed how, in the realm of security, the British retreat 'from East of Suez' in the 1960s and, in the 1990s and 2000s, its involvement in the defence initiatives of the European Union, informed the character of post-war interactions. In the field of security affairs, the literature seems to suggest that Japan and the UK are now 'reconciled

---

[24] Imaizumi Yasuaki, 'The Anglo-Japanese Relationship after the Second World War', in Ian Gow, Yoichi Hirama and John Chapman (eds.), *The History of Anglo-Japanese Relations, 1600-2000, Volume III: The Military Dimension* (Basingstoke: Palgrave Macmillan, 2003), 293-303; Hugo Dobson and Kosuge Nobuko (eds.), *Japan and Britain at War and Peace* (Abingdon, OX: Routledge, 2009).

friends' with a shared commitment to multilateral forums and international coalitions, under American leadership.[25]

Viewed from the perspective of bilateral interactions, the state of Anglo-Japanese military relations is overshadowed by the 'glorious past' of the days of the alliance. Interaction in recent years was pre-eminently prompted by the need to contribute to American-led responses to trans-national security threats to the international system. Japanese and British contributions to operations in Iraq and in the Indian Ocean are two cases in point. In Iraq, from 2004 to 2006, approximately 600 Japan Ground Self-Defence Force (JGSDF) personnel operated in the Samawah area, part of the UK controlled south Iraqi territory. In the Indian Ocean, Japan maritime Self-Defence Force (JMSDF) vessels provided fuel to coalition forces, including the Royal Navy's surface assets, and used naval facilities in Diego Garcia.[26]

By stepping away from the question of the state of bilateral relations, the chapters in this book point to the functional factors lying at the foundations of a comparison between the two island nations. Circumstances in the post-alliance and post-war periods meant that Japan and the UK entertained military relations that were no longer comparable to those established in the opening decades of the twentieth century. On the other hand, the fundamental connection between an insular geography and national security has not disappeared. How does military power contribute to Japanese and British security agendas? Reviewing recent experiences in light of how geography and strategy inform political choices in Japan and the UK today might well offer an opportunity to consolidate, expand and implement forms of future military collaboration.

The question of military cooperation is particularly relevant to contemporary Japanese and British security policy. For the past two decades, reductions in defence budgets in both countries stimulated debates and policy adjustments aimed to maximize the effect of national military resources to pursue national security goals that are strikingly similar. Both countries embarked on three major defence reviews, in

---

[25] On military interactions, cf. Yasuaki Imaizumi, 'The Anglo-Japanese Relationship after the Second World War', 299-302; Reinhard Drifte, 'Britain, Japan, and the European Union: Prospects for Regional Cooperation', *RUSI Journal*, Vol. 146, 2001:4, 83-86; Reinhard Drifte, 'Japan-UK Relations in the Global Context', in Hugo Dobson and Kosuge Nobuko (eds.), Japan and Britain at War and Peace (Abingdon, OX: Routledge, 2009), 166-177.

[26] Drifte, 'Japan-UK Relations in the Global Context', 170-171.

the mid-1990s, in the early 2000s, and again in 2010.[27] Whether under the guise of questions concerning 'normalcy' or on the nation's role in world politics, elites in Japan and the UK have been similarly concerned as to the ways in which national military power can adequately meet the requirements of the country's international profile.[28]

In the UK, the reforms of the 1990s oversaw a military transformation to create more flexible, readily deployable, expeditionary armed forces. The policies of the first decade of the 2000s, on the other hand, were implemented against fears of an increasing gap between political aspirations and the material means to match them as a result of long-term military commitments in Iraq and Afghanistan.[29] In Japan, the last decade of the twentieth century brought about a major debate on the introduction of expeditionary capabilities to favour Japanese contributions to international stability.[30] In the subsequent decade, the reconfiguration of the East Asian regional power balance, largely unfolding from the enhancement of Chinese military power, resulted in the elite's concerns over the country's ability to possess appropriate military resources for regional and international commitments.[31]

---

[27] The main documents on defence policy covering this period are: *The Strategic Defence Review*, Cm 3999 (London: The Stationery Office – TSO, 1998); *The Strategic Defence Review: A New Chapter*, Cm 5566 (London: TSO, 2002); *Securing Britain in an Age of Uncertainty: The Strategic Defence and Security Review*, Cm 7948 (TSO, 2010). In the post-Cold War era, Japanese defence policy was re-articulated in the following three documents: *National Defense Program Outline in and after FY 1996* (Tokyo: Ministry of Foreign Affairs, December 1995); *National Defense Program Guidelines in and after FY 2005* (Tokyo: Japan Defence Agency, December 2004); *National Defense Program Guidelines in and after FY 2011* (Tokyo: Japan Defence Agency, December 2010).

[28] For an overview of the debate on Japan as a 'normal country', Yoshihide Soeya, Masayuki Tadokoro, and David A. Welch (eds.), *Japan as a Normal Country? A Nation in Search of its Place in the World* (Toronto, CA: Toronto University Press, 2011). In the case of the UK, cf. Michael Clarke, *The British Defence and Security Election Survey* (Occasional Paper, London: Royal United Services Institute – RUSI, 2010); Liam Fox, 'The Strategic Defence and Security Review: A Conservative View of Defence and Future Challenges', *RUSI Lecture* (London: RUSI, 8 February 2010), http://www.rusi.org/events/ref:E4B62C2FEC5252, accessed on 10 September 2011. For a comprehensive overview on the evolution of British defence policy, see Lawrence Freedman, *The Politics of British Defence, 1979-1998* (London: Macmillan, 1999); and Matt Uttley and Andrew Dorman (eds.), *The British Way in Warfare* (Special Edition of Defence and Security Analysis, Vol. 23, 2007:2).

[29] Timothy Edmunds, 'The Defence Dilemma in Britain', *International Affairs*, Vol. 86, 2010:2, 378-382.

[30] Alessio Patalano, 'Japan's Maritime Past, Presence and Future', in Till and Bratton, *Sea Power and the Asia-Pacific*, 104.

[31] Author's interviews with senior civilian officials, Tokyo, Japan Ministry of Defence, and with senior military officers, Tokyo, Japan Maritime Self-Defence Force Staff College, 20 September 2011. Also, Patalano, 'Japan's Maritime Strategy', 85-87.

In both countries, these debates are leading to a reassessment of the costs and benefits of the security relationship with the United States. Equally similar is the debate concerning the impact of politics and bureaucracy on the development of appropriate national security strategies.[32]

In reviewing the Japanese and British experiences, this book seeks to identify areas of common security and operational interest that could help develop stronger military ties and reduce the impact of the limitations of national resources. In this respect, military cooperation is regarded as a force multiplier in the achievement of Japanese and British security objectives respectively. This book is not about the revival of Anglo-Japanese relations as a way to bring a glorious past back. It is about exploring the strategic considerations that made the security policy of the two 'Islands Empires' relevant to each other, and assess if closer military ties still make strategic sense to the security of two 'maritime nations'.

The time is certainly ripe for such cooperative actions to be explored. For one, Japanese security policy, notwithstanding constitutional limitations on the use of military power, is very different from the Cold War. As two scholars recently argued, Japan is now entering the 'fifth phase' of its post-war security policy, one in which Japan is seeking to consolidate its role as a 'global ordinary power' on the UK model.[33] In this regard, in the opening remarks of a ceremony recently held in London to celebrate the establishment of the Japan Self-Defence Forces (JSDF), the Ambassador of Japan to the UK pointed out that today the two countries 'enjoy excellent relations' and are 'close allies and partners in many areas including counter-terrorism, free trade, (and) international development'. He further reminded his guests that a little more than a century ago another Japanese diplomat with his same family name, Viscount Hayashi Tadasu, had signed the first Anglo-Japanese Alliance. In drawing this reference, he specified that he did not wish to suggest 'that Japan and the UK should seek again to conclude such an arrangement with a view to war'. Rather, he hoped for 'more vigorous interaction and

---

[32] Soeya, Tadokoro, and Welch, *Japan as a Normal Country?*, 5-6; Paul Cornish and Andrew M. Dorman, 'Breaking the Mould: The United Kingdom Strategic Defence Review 2010', *International Affairs*, Vol. 86, 2010:2, 395-410.

[33] Takashi Inoguchi and Paul Bacon, 'Japan's Emerging Role as a "Global Ordinary Power"', *International Relations of the Asia-Pacific*, Vol. 6, 2006:1, 2.

cooperation' in defence and security matters, 'a relationship amounting to a sort of alliance, this time for global peace'.[34]

In the UK, the current coalition government expressed its willingness to enhance bilateral relations. Last year, shortly after taking office, William Hague, the Foreign Secretary, stressed the significance of bilateral relations between Japan and the UK in a visit to Japan to present new directions in British foreign policy. 'Japan matters to Britain' as he put it, adding that the country is the UK's 'closest partner in Asia'.[35] Whilst his remarks pre-eminently referred to the two countries' economic ties, it is undeniable that the reconfiguration of British military power set forth by the 2010 Strategic Defence and Security Review will require an attitude aiming to strengthen international partnerships to address the plethora of security challenges to the international community, especially if the UK is to retain its primary role in world politics.[36]

For all these reasons, the book sets out to debate how Japan and the UK share today a much closer military profile than ever in the past; they are maritime nations with relatively limited military capabilities but global economic interests. More than ever in the past, both countries are connected by the necessity to address how best to employ their respective military power to fulfil the requirements of both national defence and international stability.

The Structure of the Book

The chapters that follow are organized in three parts corresponding to the pre-war, Cold War, and post-Cold War eras, respectively. The book follows a common methodological approach informed by a high degree of analytical eclecticism with the aim to take full advantage of the different academic background and professional expertise of the contributors. Earlier chapters tend to be methodologically closer to fields of

---

[34] Opening remarks by H.E. Hayashi Keiichi, Ambassador of Japan to the UK at the Japan Self-Defence Forces Day Reception, London, 30 June 2011.

[35] William Hague, 'Britain's Prosperity in a Networked World', Tokyo, 15 July 2010, http://www.fco.gov.uk/en/news/latest-news/?view=Speech8id=22540543, accessed on 4 July 2011.

[36] In 2010, the signing of an Anglo-French defence cooperation – which will lead the two countries to share military resources such as fighter aircraft, aircraft carriers, and special forces – has been regarded by commentators in the UK as a political choice dictated by defence cuts. Cf. Paul Reynolds, 'The "South-Atlantic Question" in French-British Plan', *BBC News*, 2 November 2010, http://www.bbc.co.uk/news/uk-politics-11666320, accessed on 10 January 2011.

military and diplomatic history; those in the last section reflect more the agenda and methodology of political science and security studies.

Part I of the book addresses the role of the sea in the development and termination of an Anglo-Japanese strategic partnership at the beginning of the twentieth century. John Ferris explores this role, suggesting that the Pacific Ocean represented the 'fulcrum' of power competition in Asia. His chapter presents a comprehensive picture of the Pacific as a strategic theatre, pointing out how different British and Japanese industrial and economic bases were critical in the way the two countries evolved from strategic partners to competitors on the verge of the Second World War. Haruo Tohmatsu complements this overall picture with a Japanese perspective on the importance of sea power in the country's approach to the alliance with Britain. Tohmatsu's chapter highlights the degree to which the strategic use of the sea enabled Japanese elites to choose the country's degree of involvement in international affairs at the time of the First World War, serving as a primary tool of statecraft in the pursuit of national security objectives. The final contribution to this first part is offered by Douglas Ford who focuses on an analysis of the British strategy to defend Singapore in the years preceding the war. In the chapter, it becomes apparent how intelligence assessments of Japanese intentions put a premium on a maritime strategy with Singapore as the main logistical hub for the defence of British imperial interests in East Asia. In this context, the neutralization of the naval base represented a necessary step for Imperial Japan to secure regional dominance.

Part II investigates Japanese and British defence and security policies throughout the post-war period, approximately from the opening stages of the Cold War to the 2003 intervention in Iraq. In the first chapter of this section, General Yamaguchi reflects on his experience in the JGSDF detailing the evolution of the missions and capabilities of the Japanese military. He identifies the shifts in military posture, stressing how the JSDF had to evolve from a force designed for the defence of the Japanese archipelago to a more flexible and expeditionary-oriented one, engaged in a plethora of conventional and trans-national challenges. In the subsequent chapter, Eric Grove pens the role of sea power in British post-war defence policy. He examines the specific case of the Royal Navy, analysing the changes undergone in the country's use of naval power from its primary application in ASW operations to defend Western commerce from raiding Soviet assets to a force enabler for the projection of military power in distant theatres. Finally, Chiyuki Aoi draws core themes approached in the previous two chapters together by analysing the impact of Japan's more recent experience in stability operations on the ongoing transformation of the country's defence policy. The chapter clarifies that what is commonly labelled as 'Japan's attempt to acquire power projection capabilities' is in reality a much more complex adjustment aimed at acquiring capabilities to contribute to expeditionary missions in *ad hoc* formulas.

The final part of the book focuses on matters of current national security policy concern and maritime strategy in Japan and the UK. French specialist Guibourg Delamotte sets the tone of this section engaging with the controversial issue of the impact of legal constraints on political action in Japanese security. Her chapter suggests that these self-imposed limitations have had repercussions on the procurement of military capabilities and commitments to operations on land. Yet, a more in-depth analysis of Japanese post-Cold War international operations would equally suggest that the boundaries of these constraints have been pushed by naval activities, which in turn helped renegotiate the political and legal acceptability of the scope of Japanese international interventions. Commodore Jermy's chapter turns the attention back to the UK, pointing to the heart of some of the most cogent issues affecting British defence and security policies, notably current operations and strategy in Afghanistan. The chapter explains how for the UK it is essential to maintain expeditionary capabilities, notably amphibious assault ships and aircraft carriers. These assets are, his chapter argues, functional to the country's ability to influence events on land and operate to secure stability in areas of the world vital to national security. Admiral Kōda complements the ideas presented in the previous chapter. His chapter stresses how, in the daily operational activity of Japanese assets deployed overseas, cooperation with the UK is an invaluable asset. More than six decades of military partnership with the United States have made the two countries' armed forces naturally inter-operable. The chapter further highlights that both the UK and Japan are involved in areas like the Gulf of Aden, the Indian Ocean, the Horn of Africa, and previously Iraq, based on shared interests in maintaining a close partnership with the United States, as well as stability and security in areas of passage for shipping carrying energy supplies and goods of mass consumption.

Looking ahead, it is undeniable that Japanese and British military and security policies are undergoing important transformations. Japan's military profile is growing in importance regionally as well as internationally. A comparative analysis with the UK, a country with similar geographic features and international standing, offers a valuable reference on the challenges that Japanese policy-makers will face as this transformation continues and questions on how to best deploy military power emerge. For the UK, such an approach is equally relevant for it underscores the extent to which, in the twenty-first century, the maintenance of an effective and capable expeditionary force is a national strategic imperative, and *ad hoc* partnerships with like-minded partners are a potentially crucial means to increase the effectiveness of national military resources.

PART ONE

STRATEGIC PARTNERSHIP AND MILITARY
RIVALRY ACROSS THE OCEANS

CHAPTER ONE

THE FULCRUM OF POWER: BRITAIN, JAPAN AND THE ASIA-PACIFIC REGION, 1880-1945

John Ferris

This chapter addresses power, especially a struggle over it, in the Asia-Pacific region. Until the nineteenth century, the maritime theatres included in this vast area were barely connected. Power in Asia was diffuse, scattered between regions, which largely were self-contained. No one dominated the Pacific through power, until the United States Navy (USN) crushed the Imperial Japanese Navy (IJN) in 1945. The first 'sea power' in the Pacific Ocean, the Polynesians, used it to colonize. The second, Spain, did so for commerce. Polynesians worked without warships, Spain with few of them. So too, in the nineteenth century Britain dominated the Pacific region through merchants, bases, and the potential for presence. It could barely control clans on Pacific islands though, when challenged by any rival, as in the case of the United States around Oregon between 1844 and 1852, Britain could deploy sufficient warships to protect its interests.[1] Britain dominated the Pacific Ocean not through its strength, but the absence of rivals. By mid-nineteenth century, Britain (and Russia) acquired greater power over Asia than anyone had done since Tamerlane. In the process, it oversaw the creation of a new economic and strategic ecosystem, the Asia-Pacific region, which became central to the architecture of its empire, and the world.

This structure, its centrality and vulnerability, framed the Anglo-Japanese struggle for power. In turn, that struggle was a bilateral relationship in a multilateral context. Third parties drove it as much as anything the principals did. Thus, around 1900, Russia thrust Britain and Japan together, when each preferred a deal with it; the rise and fall of Chinese governments pressed them together, and apart. This external pressure twisted Anglo-Japanese relations in powerful and

[1] Barry Gough, *The Royal Navy and the Northwest Coast of North America, 1810-1914: A Study of British Maritime Ascendency* (Vancouver, CA: University of British Columbia Press, 1971); Jane Samson, *Imperial Benevolence: Making British Authority in the Pacific Islands* (Honolulu, HI: University of Hawaii Press, 1998).

unpredictable ways, because East Asia was the most fragile part of the international system, affected by every development abroad, and shaken by constant changes in every country's policy and the grinding of the tectonic plates of power. The actors were not just states, but people, and peoples. They pursued interests, which combined things with ideas about them: imagined objects or objectified images. States exercised power through leverage, presence, deterrence and war. Power had parts: armed force, economic strength, and the political ability to set rules, or to coerce or attract other states. These elements were linked dialectically, not just in a linear mode. Economics created sea power. Trade followed the flag: force created economic regimes, while states struggled to control international systems for national advantage. To set rules for the Asia-Pacific region and to dominate it strategically, had economic causes and consequences. To create wealth, or take it, made power.

In this struggle, Britain fell victim to its own success. It guided the creation of an ecosystem, which had economic and strategic niches anyone could exploit for their purposes, and generated wealth enough to strengthen rivals, in ways no one could anticipate. Preeminent among these unpredictable consequences was the rise of Japan against Britain. This change in power is not easy to measure. Thus, arguments of economic determinism do so through simplistic means. In this story, Britain starts high, but declines steadily toward imperial overstretch, driven by the rise and fall of its economy, while Japan follows the opposite course. In fact, British decline began late in the day, as did its Japanese problem. Economic developments were not the, or the only, fundamental cause for these phenomena. Though Japan grew steadily, Britain did not develop or decline that way. On several occasions it jumped suddenly from one quantum level to another, up or down. The competition between these states centred on a particular combination of those elements of power: maritime strength, the industrial base for it, and dominance in the Asia-Pacific region.

Sea Power and British Economic Interests
in Nineteenth Century East Asia

During the nineteenth century, a combination of the weakness of local states and the strength of commercial desires, whether expressed by businesses or anticipated by statesmen, about existing or imagined objects, drove British expansion in Asia and the Pacific. Much of its direction was unofficial. Most of its force was commercial, emanating from individuals and firms which often were not British, or even Western, but found cooperation beneficial. Its military edge – large forces in India and small fleets in East Asian waters, sometimes augmented by small, expeditionary forces – overwhelmed local competition. Britain received a high strategic and economic return on that investment. It used force as a tool of capitalism and liberalism: to open markets, to define terms of trade, to turn ambitions into interests, to anchor allies and to change minds. Both from ideology and to make allies, unlike previous colonial ventures of Europeans in Asia, Britain created an open rather than a mercantilist system, the largest tariff free zone on earth, or in history.

In the 1860s, British fleets dispatched to coerce China into that system did the same to Japan. The exercise of power, through an international coalition and actions such as destroying coastal fortifications, taught Japanese leaders that the foreigners could not be driven out and should not be dragged further in. One British diplomat noted in 1865, "the presence of powerful material support and the determination to be ready to rely upon it, as a last resort, will for some years to come be the basis on which a successful representation of Her Majesty's interests and the real happiness of Japan itself must depend".[2] Japan accepted unequal treaties, losing control over tariffs or jurisdiction over foreigners, and entered a quasi-colonial status, between independence and protectorate. For the next generation, Japan's strongest relationship was with Britain. The latter, secure in power and certitude, paid little attention to Japan but also did not try to eliminate its independence. Britain was willing to let Japanese succeed economically and politically, if they played by British rules.[3] Japan had to learn the British game, in which it had the edge of the underdog. Britain had to play all comers, among which Japan was too peripheral to warrant special tactics. By evolving to compete in an ecosystem created by Britain, and learning to exploit

---

[2] British Parliamentary Papers, *Reports and Correspondence Concerning Japan, Volume Two, 1864-1870* (Shannon, IR: Irish University Press, 1970), 405, Winchester to Russell, 26 May 1865.

[3] Grace Fox, *Britain and Japan, 1858-1883* (Oxford: Oxford University Press, 1969); Olive Checkland, *Britain's Encounter with Meiji Japan, 1868-1912* (Basingstoke: Palgrave Macmillan, 1989); James Hoare, 'The Era of the Unequal Treaties, 1858-99', in Ian Nish and Yoichi Kibata (eds), *The History of Anglo-Japanese Relations, 1600-2000, Volume I, The Political-Diplomatic Dimension, 1600-1930* (Basingstoke: Palgrave Macmillan, 2000), 107-30.

its weaknesses, Japan found comparative advantages and a niche of its own. By the time Britain realized the significance of these develop-ments, they were hard to overcome. British policy towards Japan was marked by some foresight, more than any other state exhibited on the matter, but its leaders could not see sixty years ahead. Britons did not treat Japan as a central or coherent matter. Japanese did better in bar-gaining, often manipulating confusion, sometimes creating it.

The balance between Britain and Japan first shifted in the 1880s. Japan, perhaps more than any other state, profited from access to the free trade system which Britain paid to maintain across the world, because its home market was so small. Japan was legally limited in its abilities to exploit that system, by adopting a mercantilist approach and openly protecting industries, as western countries routinely did, but still the state applied all the subsidies it could, and worked closely with firms to pursue common aims. Together, they used British models and technology, which Britain willingly sold to all comers, to develop sec-tors able to compete with western exports, mostly English and con-sumer goods. By 1880, Japanese firms surpassed British ones in the cost and quality of exports to their islands, and increasingly in nearby markets. Britons, official and unofficial, emphasized Japan's remarkable economic and civil progress.[4]

As Japan prospered, British trade stagnated. Businessmen based in the country, the sectional interest which hitherto had dominated British policy there, blamed this failure on Japan, and demanded that the treaties be better enforced, or rewritten in their favour. This aim was impossible, given the legal need to renegotiate the treaties and Japanese success in splitting western powers over the matter. By the 1890s, diplomats and Whitehall concluded that this situation could improve only by accepting Japanese demands to abolish the unequal treaties – the first break from extra-territoriality, its standard approach towards non-western countries – and by pursuing a new relation-ship, based on equality in form. The British legation in Tokyo agreed that British firms must suffer from Japanese tariffs, but hoped Britain 'would reap the benefit of our liberality' by establishing 'a claim to the

---

[4] Janet Hunter and S. Sugiyama, 'Anglo-Japanese Economic Relations in Historical Perspective, 1600-2000: Trade and Industry, Finance, Technology and the Industrial Challenge', in Janet Hunter and S. Sugiyama (eds), *The History of Anglo-Japanese Rela-tions, 1600-2000, Volume IV, Economic and Business Relations* (Basingstoke: Palgrave Macmillan, 2002), 1-109.

goodwill of a nation whose future contains possibilities which cannot be ignored'.[5] The freedom to raise tariffs improved Japan's abilities to exploit the British economic system. Still, the legation predicted that Britain could profit from the great industrial development that was in progress throughout the Empire, and

> the prospect of an unlimited demand for British manufactures now offered, not only by the Government for war material, bought by the numerous joint stock companies now being successfully floated for docks, railways, factories, etc... (and) the demands of the people in general for many articles of daily use which Great Britain can supply a demand which is steadily growing with increased wealth and ideas of comfort and necessity.[6]

Britain and the Emergence of Japanese Sea Power

Meanwhile, Japanese leaders, haunted by vulnerability, pursued security and power through the development of armed forces modeled on the best western institutions. Throughout the 1870s and 1880s, British sailors trained the Imperial Japanese Navy (IJN). Britain routinely offered smaller nations such services, but Japan profited from them more than any other country. It developed a good and medium sized navy, with equipment built abroad, mostly in Britain. This experience gave the Royal Navy (RN) an accurate estimate of the quality of the IJN which, by 1894, it ranked far above the fleets of China (or any other non-western state) and level with many western powers. British military authorities had the same, accurate, view of the Imperial Japanese Army (IJA).[7] Over the next forty years, Whitehall assessed Japanese capabilities well, better than other powers did.[8] British policy also remained consistent with its interests and estimates.

---

[5] United Kingdom National Archives, London (UKNA), FO 811/5773, Memorandum by Gubbins, 3 January 1888. Yu Suzuki, 'Modern Japan, Episode Zero, Japan's Struggle for Diplomatic Equality, 1859-1894' (Unpublished MA dissertation, The University of Calgary, 2010).

[6] UKNA, FO 46/466, British Legation, Tokyo to Foreign Office, 14 January 1896.

[7] UKNA, FO 881/6594, Admiralty to Foreign Office, 16 July 1894, Intelligence Division to Foreign Office, 16 July 1894.

[8] John Ferris, 'Turning Japanese? British Observers of the Russo-Japanese War', in John Chapman and Inaba Chiharu (eds), *Rethinking the Russo-Japanese War, 1904-5, Volume II, The Nichinan Papers* (Folkstone: Global Oriental, 2007), 119-34; see also Ferris, 'Japan In The Eyes of the British Army and the RAF, 1919-41', in Ian Gow, Yoichi Hirama and John Chapman (eds), *The History of Anglo-Japanese Relations, 1600-2000, Volume 3 The Military Dimension* (Basingstoke: Palgrave Macmillan, 2003), 109-26.

By 1890, Japan became a second class power in East Asia, and began to affect British policy and international relations. The outcome was not preordained. Its first emanation was in the Sino-Japanese War of 1894, where British policy stemmed partly from personality. As Japanese actions threatened British commercial interests, and political ones, by destabilizing China, the Foreign Office wished to intervene so as to stop Japan. The Prime Minister, Lord Rosebery, however, treated Japan as a power with independent rights, and some cause in this case, reinforcing his usual caution. He adopted a neutral position which favoured Tokyo during the war, and after it, by refusing to join the Franco-German-Russian intervention that made Japan disgorge some gains. By the war's end, Rosebery rated Japanese strength highly, and believed it might matter in power politics, by becoming a check on Russia. His Foreign Secretary, Lord Kimberley, pursued cooperation with Japan against Russia. The next Prime Minister, Lord Salisbury, less impressed by Japan, still recognized it as significant in regional politics, sharing interests with Britain against Russia.[9] His preference, like that of all British statesmen, was an arrangement with Russia. Nor did Japanese leaders wish to fight Russia, unless it made itself a threat. Japan and Britain were driven together only by Tsarist belligerence and recklessness, its disregard of their interests, combined with the implosion of the Ch'ing state during the Boxer Rebellion, and Russia's ability to dominate Beijing, and Korea. Had Rosebery or Russia acted otherwise, Anglo-Japanese relations would have taken a different road.

The Anglo-Japanese Alliance of 1902 was not a love match, but a marriage of convenience, with secrets kept on both sides. For Britain, the arrangement primarily was diplomatic, and for Japan, strategic. Britain acted to restrain Russia and Japan – to prevent Tokyo from becoming a Russian pawn, through diplomacy or defeat, and to solidify the *status quo* in Asia. Although it met the letter of the alliance, Britain helped Japan less during the Russo-Japanese War than Whitehall later liked to pretend. Japan, conversely, used the alliance to strengthen its bargaining position against Russia in peace, while clearing the decks

---

[9] National Library of Scotland, Edinburgh, MS 10243, Rosebery to Kimberley, 10 April 1895, Lord Kimberley Papers; Keith Neilson, *Britain and the Last Tsar, British Policy and Russia, 1894-1917* (Oxford: Oxford University Press, 1996); Thomas Otte, *The China Question: Great Power Rivalry and British Isolation, 1894-1905* (Oxford: Oxford University Press, 2007).

for war – to force a Russian rollback from Korea, one way or another, without informing Whitehall of its intentions.[10]

The Russo-Japanese War made Japan a great power in Asia. The result for Britain was unexpected, and too much of a good thing. The war solved one problem in Asia, while creating another, and destabilizing Europe. It also raised the question of what to do with the Anglo-Japanese Alliance, as the original rationale was history and revision necessary. After 1905, the alliance became a strategic relationship for both sides, a reinsurance treaty marked by reservation and bound by inertia. Each state aimed, as much as anything else, to contain and maintain certainty about the other. Breaking up was hard to do, harder than staying together, especially because it would force each state to ask who would be its – and the other's – new partner. Weak external forces also held the allies in harness. The possibility of Russian hostility concerned them, barely, the German problem pressed Britain on Japan, as the American one moved Tokyo towards London.[11]

Beneath this surface were fractures. Britain's refusal to promise Japan military support against the United States divided the allies, as did differences over China. A British faction saw Japan as a problem and possibly a threat. A greater Japanese group saw Britain as an enemy in the way of an imagined empire, and as a target; Germany or Russia as allies, and China as a rationale for war.[12] Their views fit the predictions of those British officials who imagined a "yellow peril" led by Japan.[13] Japanese militarists dreamed big and overestimated their strength. They wanted objects which Britain had, yet realized they were too weak to take them, and so imagined some ally could defeat Britain or distract its strength, as their descendents would do in 1941. Signs of weakness stimulated the militarists like blood in the water. They could be deterred only by a superiority of power too great to mistake. Before 1918, however, the militarists, like the Russophobe faction in Britain, had influence, but not dominance, and faced opposition. They were

---

[10] Nish, *The Anglo-Japanese Alliance*.

[11] Ian Nish, *Alliance in Decline, A Study in Anglo-Japanese Relations, 1908-23* (London: Athlone Press, 1972); Peter Lowe, *Great Britain and Japan, 1911-1915: A Study of British Far Eastern Policy* (London: Palgrave Macmillan, 1969).

[12] Frederick Dickinson, *War and National Reinvention: Japan in the Great War, 1914-1919* (Cambridge, MA: Harvard University Press, 2001).

[13] John Ferris, 'Armaments and Allies: The Anglo-Japanese Strategic Alliance, 1911-1922', in Phillips Payson O'Brien, *The Anglo-Japanese Alliance, 1902-1922* (Abingdon, OX: Routledge, 2004), 251-2.

contained by other factions, which realized, rightly, that Britain was too strong to fight.

Through the revision of the Anglo-Japanese Alliance in 1905 and 1911, each party achieved its aims, in different ways. For Britain, renewal contained Russia and Japan, and the possibility the latter might join with Germany. The alliance saved the need for defence in the Pacific and protected interests in China, by turning a problem into a solution. These gains were moderate, as Britain's position was strong, but they eased its problems before 1914 and proved useful during the First World War. Meanwhile, economic developments justified Britain's hopes of 1888. It found a new role in the Japanese economy, as the primary influence in the development of heavy industry. Japan became a significant, if secondary, market for the most sophisticated machinery manufactured in Britain, civilian and naval.

A Shifting Balance of Power

These developments produced greater gains for Japan, as it entered a phase of militarized mercantilism, pursuing the industrial capability needed to produce weapons at the state of the art. British exports to Japan centred on the construction of warships. Between 1908 and 1918, Britain perfected the production techniques for dreadnoughts and sold them to Japan, through the greatest transfer of naval technology in history, involving the leading British armament firms, Vickers and Armstrong & Whitworth, and a Japanese business, the Hokkaido Tanko Kisen Co. The British firms provided all the information needed to establish a plant in Muroran, able to produce all forms of naval armament, especially amour and guns. The IJN sponsored this project because Japanese firms could not make the high-grade steel and large-calibre guns needed for the second generation of dreadnoughts. The factory used British technology and modes of organization, working under the supervision first of Vickers technicians and then of Japanese trained by them. Vickers offered such opportunities to many states, like Russia or Spain; Japan exploited them better.[14] This was one of the first

---

[14] Bunji Nagura, 'A Munition-Steel Company and Anglo-Japanese Relations Before and After the First World War: The Corporate Governance of the Japan Steel Works and its British Shareholders', in Hunter and Sugiyama, 'Economic Relations', in Hunter and Sugiyama (eds), *Anglo-Japanese Relations, IV*, 156-82; Kozo Yamamura, 'Japan's Deus ex Machina: Western Technology in the 1920s', *The Journal of Japanese Studies*, Vol. 12, 1986:1, 65-94.

three instances of what became Japan's standard means for industrial modernization (another was the simultaneous establishment of a firm using technology from Nobel Explosives). Firms worked with foreign concerns, producing their products under licence, aided by the parent, acquiring expertise in practices just behind the state of the art and learning to walk on their own. Often, these liaisons were essential in teaching Japanese firms new forms of engineering. Japan could not have entered the Dreadnought era without British help, which let it tie the United States in naval technology, surpass France and Russia, and raise its status at sea by pursuing a programme to construct eight battleships and eight battlecruisers.[15]

The First World War changed the balance between Britain and Japan as much as everything which had occurred in the past twenty years. The fractures in their interests increased, especially over Japanese aggrandisement in China. British officials came to see Japan not as a friend, but as a rival, though rarely as an immediate threat. Japanese militarists regarded Britain as an obstacle, and wanted to align with Russia or even Germany against it, which would have mattered had the war ended in a stalemate, as British leaders thought likely during 1916-18. Equally, after the United States entered the war, Whitehall favoured unofficial American proposals for an entente directed partly against Japan. These splits and suspicions declined after 1918, but were remembered. Meanwhile, Japanese power rose, but it remained far weaker than Britain's. During the 1920s, Britain was viewed as being stronger than was true, but strong it was, and the only world power, with the greatest maritime capabilities on earth and remarkable ones in aviation.[16] The years 1815 and 1919 marked the RN's two great peaks of power in history. Britain maintained the naval strength and industry needed to handle a two power standard of Japan and any European power, and to beat Japan in a naval war. These capabilities, however, declined sharply around 1930, as much as they did between 1914 and 1918, through a combination of developments in economics, power politics and strategy, as problems scattered across the range of power became a cascade. Not merely would Japan become stronger; Britain would become weaker.

During the First World War, Japanese capacity in key areas of industry like shipbuilding doubled, through the stimulus of foreign orders,

---

[15] Ferris, 'Armaments and Allies', 259-60.
[16] John Ferris, '"The Greatest World Empire on Earth": Britain During the 1920s', *International History Review*, Vol. XIII, 1991:4, 723-47.

especially British ones. For Japan, the British economy became less important, though still a clear second behind the United States. British officials and industrialists increasingly recognized Japan as competition, as well as complement. By 1917, the Board of Trade's expert on East Asia, Thomas Ainscough, saw Japan as a 'scientific, developed, highly-protected industrial nation', a powerful competitor to Britain, able to assimilate modern technology and a good market, but only for the most advanced equipment and the means to produce it. 'The greatest opportunities would appear to be in machinery and plant for the equipment of the new factories of all kinds which will be erected during the next few years'. Yet, Ainscough indicated that Japanese firms would use such opportunities to modernize their industrial base and compete with Britain in new markets and products.[17] Though he soon concluded that Japan had failed to exploit these opportunities, from 1925 his prophecies came true regarding textiles and, in different spheres, warships and aircraft. In 1921, when considering a Japanese commercial mission to Britain, the Department of Trade held that because of their 'unscrupulous methods of copying processes etc. when privileges are extended to them', Britain should not show 'competent Japanese engineers over our works. After all we are one of the foremost industrial nations of the world & in general have much more to lose than to gain by any reciprocal action'.[18] By 1930, while part of a commercial mission visiting Japan, one British industrialist wrote, 'I have not heard of any English firms out here making progress (oil & talking machines excepted) – I do hear some of them living in the hope that their work here may last their days'.[19] In a break from the orthodoxy of free trade, by 1930 the British empire used tariffs to protect Britain's textile industry from being devoured by its Japanese competitors.[20]

These developments mark the trend from 1860 to 1940, but they merit context. Japan profited against Britain from one sided advantages: official and unofficial tariffs, technology transfer, and piracy of patents. Its textile, shipbuilding and steel industries, the first modern ones, stemmed from British models, with access and advice freely granted.

---

[17] Board of Trade, 'Report on the Probable Position of British Trade in the Far East After the War', by Thomas Ainscough, 15 February 1917, Steel Maitland Papers, GD 193/72/2, National Archives of Scotland, Edinburgh.

[18] Minutes by MacKinnon and Paish, 19 November 1921 , BT 60/3/2.

[19] Capital Steel Works, Sheffield, 11 April 1930 to Edward Crowe, BT 60/8/2.

[20] Hunter and Sugiyama, 'Economic Relations', in Hunter and Sugiyama (eds), *Anglo-Japanese Relations, IV*, 53-6.

No Japanese industry could compete against British ones without the aid of tariffs until 1927, when its firms developed superiority over those in one staple industry, textiles, and made inroads into another, shipping. Before 1945, British firms remained superior in costs and quality to Japanese ones in all sectors of heavy industry. In 1938, the Japanese economy was far smaller, narrower and less advanced than that of Britain. Yet, Japan had jumped far into the industrial age, with Britain's help and at its expense.[21] Between 1910 and 1930, especially during the later 1920s, suddenly, for the first time, an industrial base emerged in the northwest Pacific, thousands of kilometres from its nearest competition, well placed to exploit the Asia Pacific region, and to challenge British dominance. While not in the first rank of industrialized economies, the United States, Britain or Germany, Japan was on par with those of the second class, France, Italy, and the Soviet Union, and rising.

In particular, Japan's warship and aircraft industrial sectors had fair breadth and depth. Its maritime forces and industries were a carbon copy of British models, blurred in places, but with time available to sharpen the focus. Just as before the war, after 1919 the IJN and Japanese firms pursued British help in areas where they lacked expertise, especially naval aviation. The Admiralty tried to block such help, because it regarded Japan as a threat. Japan evaded that embargo, by exploiting informal ties which existed under the alliance, especially the eagerness of the Royal Air Force (RAF) and British businesses to find markets in Japan. Britain propelled Japan into the air age. British aviation firms were fundamental to the establishment of aircraft manufacturing in Japan. An 'unofficial' air mission gave the IJN a sound footing in naval aviation, based on the doctrine and practice of the world's leader in that arm. While Britain attempted to prevent Japan from gaining expertise with aircraft carriers, the IJN sidestepped this blockade by hiring the British traitor, Frederick Rutland, the officer in the world with the most experience in organizing aircraft on such vessels.[22] Meanwhile, poisonous relations between the RAF and the RN lost Britain its lead in naval aviation.

British designs, patents and equipment were essential to every Japanese warship of 1921, and to their firepower – to the basis of the

---

[21] For British and Japanese economic performance, cf. John Sharkey, 'British Perceptions of Japanese Economic Development in the 1920s, with Special Reference to the Cotton Industry' in Hunter and Sugiyama (eds), *Anglo-Japanese Relations, IV*, 249-82.

[22] Ferris, 'Armaments and Allies', 253-4, 260-1.

IJN of the interwar years, and the Pacific War. Britain was not Japan's only potential source for aid in warship building and naval aviation, just the best. No other country matched its quality in these spheres. British firms gained from these actions, but Japan even more. The IJN combined British material and organization with Japanese men, spirit and tactics, developed these capabilities with independence and excellence, and applied them to ends of their own. As Britain divorced Japan, it paid alimony.

Between 1914 and 1918, the multilateral context to Anglo-Japanese relations changed, as it would again around 1930 and 1940.[23] In systematic terms, by 1919, the world system was fragile, especially in eastern Europe and China and as regards international finance, but bolstered by armed liberalism. Britain, the United States and France, fairly united and heavily armed, had overwhelming power and loosely cooperated to support the *status quo*. Their diplomacy clashed, but their strategy coincided. Masters of the seas and Europe, their strength bound the fractures in the world order, providing time to heal, deterring threats and attracting allies. The revisionist powers, weak and isolated, had to play by liberal rules, because they were backed by power. These conditions weakened the militarists in Japan and strengthened their opponents, making their country a junior member of the liberal coalition.

The East Asian Order in the Interwar Period

Other particular developments framed Anglo-Japanese relations, again binding fractures beneath the surface. The German threat to Britain, and opportunity to Japan, was broken for fifteen years. Russia, down but not out, bound Britain and Japan more in its Soviet form than the Tsarist version had done between 1905 and 1914. The disintegration of the Chinese state, followed by its revival combined with nationalism, first divided Britain and Japan, joined them between 1922 and 1928, and then split them again. The collapse of Russia and China left Japan stronger in East Asia than ever before, and potentially dominant there.

---

[23] For Anglo-Japanese relations between 1918 and 1941, cf. the essays in Nish and Kibata, *The History of Anglo-Japanese Relations, Volume 1*; and *Volume 2*, and Anthony Best, *Britain, Japan and Pearl Harbor: Avoiding War in East Asia, 1934-41* (Abingdon, OX: Routledge, 1995).

Equally, the collapse of Russia and Germany destroyed the external powers with which Japanese militarists had hoped to form an imagined coalition, and produced the Soviet Union, which they loathed.

These circumstances left Japan alone against Britain, the United States, or both. Even the militarists could not mistake this disparity in strength. Finally, the United States became potentially the world's greatest power, affecting Britain, Japan and their relationship in dynamic ways. Japanese perceptions of, and policy towards, Britain became linked to those regarding the United States, to matters beyond British control. Japanese leaders were divided between those willing to accept Anglo-American rules and superiority in strength, and those wanting to treat the United States as a threat. So too, in 1917-19, British statesmen were willing to accept an Anglo-American condominium across the world, a possibility which vanished because of confusion in Washington about what to do with its power. Instead, until 1942, the United States and the United Kingdom weakened each other. They shared interests, and wanted the same world, but differed over who would do what to whom within it: American leaders rejected British superiority, but claimed parity, which meant dominance, in the relationship. Only one problem limited their ambition; American leaders would not pay any price to achieve these aims or to realize their potential. They could achieve their aim only by cutting Britain down to size – their size. In doing so, American leaders damaged British power, and the international system, in ways they did not understand, or intend.

The first phase in fratricide occurred by 1921-22, with the Washington Conference, which produced a 5-5-3 ratio in battleships and aircraft carriers between the British, American and Japanese navies, a withdrawal of Japanese forces from the parts of Siberia and China it had occupied during the war, and the end of the Anglo-Japanese Alliance.[24] American pressure shaped two of these outcomes, less so the last. British perceptions of Japan had turned. Politicians and the Foreign Office hoped to retain the alliance as a loose reinsurance treaty.

---

[24] On this subject, see Ian Nish, Alliance in Decline: Study in Anglo-Japanese Relations, 1908-1923 (London: Athlone Press, 1972); John Ferris "'It is our Business in the Navy to Command the Seas": The Last Decade of British Maritime Supremacy, 1919-1929', in Keith Neilson and Greg Kennedy (eds), Far Flung Lines: Maritime Essays in Honour of Donald Schurman (London: Frank Cass, 1996); Erik Goldstein and John Maurer (eds), The Washington Conference, 1921-1922: Naval Rivalry, East Asian Stability and the Road to Pearl Harbor (Abingdon, OX: Routledge, 1994); Sadao Asada, 'From Washington to London: The Imperial Japanese Navy and the politics of naval limitation, 1921 1930', Diplomacy & Statecraft, 1993:4, 147-91.

The Dominions, however, rejected its continuation, while the Admiralty and the War Office opposed reliance on Japan and believed they could handle the problem. These events had mixed results. For the first time, Britain confronted a threat in the Asia-Pacific region, which it had to match through power, but the RN retained a strong position to do so. During the 1920s, British policies toward Japan were aligned, involving diplomatic cooperation alongside strategic precautions, like naval rearmament and the construction of the Singapore base. While the RN defined Japan as its primary threat, essentially for budgetary reasons, the IJN did not regard Britain as hostile, or develop war plans against it, and the Japanese government cooperated with Whitehall.[25]

Underlying this success lay four problems. The RN oversold the idea of a Japanese menace, which created opposition at home to its policy, to actions necessary for security, especially because they were occupied in the politics of how to manage the United States, and the international system. The end of the alliance weakened interest groups in Japan that had worked with Britain, speeding the rise of an anti Anglo-American faction, which seized power in the IJN and expelled its rivals by 1934.[26] As American economic power rose, its leaders were doubly determined to cut Britain down to size. Finally, decision-makers in the liberal states overestimated the stability of the world, and believed that liberal internationalist solutions – cutting military forces, especially their own – could best solve those problems which remained. Liberal internationalists attacked armed liberalism, because they misunderstood the point of power.

These four problems became one in 1930, at the London Naval Conference. Here, the Labour Prime Minister, Ramsay MacDonald, aimed to spur liberal internationalism by reducing fleets and gaining American support for disarmament in Europe. Hence, he adopted American positions on naval disarmament which, of course, were selected to weaken Britain against the United States (and, incidentally, Japan). The Washington Conference established British and American fleets of 20 and 18 battleships and battlecruisers, with 10 for Japan, and agreed that from 1930, these warships would be replaced on a defined schedule. At London, the powers decided to abandon the replacement of capital ships until 1936, while immediately scrapping

---

[25] Ferris, 'It is our Business in the Navy to Command the Seas'.
[26] Ian Gow, *Military Intervention in Prewar Japanese Politics: Admiral Kato Kanji and the 'Washington System'* (Abingdon, OX: Routledge, 2004).

five British battleships, three American ones, and the Japanese battlec-
ruiser *Hiei* (which, however, rebuilt, fought hard during 1941-42, and
ultimately was not lost to Japan). After intense debate, which enraged
the anti Anglo-American faction in the IJN, the conference extended
the 5-5-3 ratio to all other categories of warships. It allowed Japan
and Britain small construction programmes for such warships until
1936, with a larger one for the United States, which had built fewer of
them since 1922. These decisions damaged Britain disproportionately.
The scrapping of warships weakened it against Japan, the United States,
and every navy. That the RN scrapped battleships equaling those of
Italy helps to explain why the latter soon felt safe to act against Britain.
The greatest effect lay in the industrial base for seapower where, after
1919, Britain possessed the world's largest capacity, the difficulty being
to feed it. When the London Conference demonstrated that orders
would be scarce for years, British firms rapidly scrapped almost 40%
of their capacity to build warships. The Japanese and American naval
industrial bases were smaller, and received proportionately, larger
orders.[27] The coherence of British policy towards seapower and Japan
broke, as ideology redefined perceptions of both matters, and the gap
between them.

In the naval disarmament negotiations of the 1920s, competition
between the United States and Britain strengthened the weakest party
against both of them. Japan did better in naval matters at the Washington
Conference than the United States or Britain. It would have collapsed in
relative strength had rivalry continued in construction. Japan did better
on a 5-5-3 ratio. This institutionalised the situation of 1921, a moment
when Japan was at the peak of its sea power, based on the best performance it could offer between 1911 and 1921, when other powers were
handicapped. During 1920 and 1923, the ratio of strength between the
RN and the IJN declined from 200% to 167%, and, by 1931 to, approximately 155%. The London Naval Treaty left Japan master of the western
Pacific and the strongest power in East Asia, which could not have happened had it confronted pressure from Britain and the United States.
These countries accepted that position because they regarded Japan as
an aligned power, and hoped the London Conference would bolster
liberal internationalist rules in the world. Instead, liberalism unleashed
militarism. The liberal powers weakened each other and lost a junior

---

[27] Ferris, 'It is our Business in the Navy to Command the Seas'.

partner. The London Naval Treaty lit the fuse for a political explosion which blew Japan down the road to the Pacific War, beginning the process by which Japan became a revisionist power and attacked liberal internationalism, and its neighbours.[28] It also weakened Britain against Japan and every other power. What seemed like a small adjustment in naval strength actually was a jump too far, past a tipping point, taking British power down one quantum level to another.

The closest that British statesmen came to define great power status was through the concept of 'sufferance', whether a state could pursue its vital interests, regardless of the wishes or forces of foreign countries. A nation which could not do so was 'merely existing on the good will' of others. Although some interests of every country must be vulnerable to other states, any nation with vital interests so exposed that it could not follow an independent policy, ceased to be a great power. Thus, Arthur Balfour asserted that if France could 'make English seaborne trade impossible… we must henceforward count ourselves among minor powers (sic). We should live on sufferance.'[29]

Britain Declining, Japan Rising

The London Treaty pushed Britain past that point with the United States and near it with Japan. By 1932, unlike 1928, the Admiralty no longer believed that the RN could defeat Japan while protecting Britain in Europe. Only its full mobilised strength could contain Japan. This view was accurate. The reduction in RN strength also raised the power of every other fleet, especially those of Germany and Italy, and made them count in any calculation of sufferance. Germany was in reach of the 'risk fleet' Tirpitz had pursued in vain, and Britain had broken at such sacrifice. Any threat in Europe would paralyse Britain in the Pacific, or vice versa – the decline of the two power standard bequeathed a two power and a two hemisphere problem. Japan became a problem not just in Asia, but in Europe. British vulnerability to a possible coalition encouraged Japanese belligerence. Britain was hostage to Japanese policy. Any problem with Japan would trigger others elsewhere, or vice versa. To maintain security in Europe, Britain had

---

[28] James William Morley (ed.), *Japan Erupts: The London Naval Conference and the Manchurian Incident, 1928-1932* (Columbia, NY: Columbia University Press, 1984).

[29] John Ferris, *Men, Money and Diplomacy, The Evolution of British Strategic Foreign Policy, 1919-1926* (Ithaca, NY: Cornell University Press, 1989), 60-1.

first to rebuild superiority over the IJN, which would take years, especially because it had to construct not just warships, but its shipbuilding capacity. Until then, Britain would remain vulnerable to a Japanese ability to attack vital interests, dependent on American aid and exposed to any European threat, which this very situation would encourage.

Such threats abounded after 1929-31, the turning point in the interwar years. The great depression wrecked the international financial system, shook the political order, and triggered crises across the globe. Armed liberalism vanished as economic and political issues split Britain, France and the United States, stripping power from the *status quo* and liberating the revisionists. Germany, Italy and Japan became greater threats. For Britain, every diplomatic relationship deteriorated, simultaneously, good becoming mediocre, mixed becoming poor. Political chaos was multiplied by economic disarray, which led even Britain to abandon free trade, to pursue a mercantilist policy that damaged every other state, for the reasons that had let them exploit it for a century. Britain confronted problems everywhere, and found solutions difficult to execute.

Special problems emerged in East Asia, where every relationship pressed Britain and Japan against each other. Chinese nationalism, and different strategies for dealing with it, divided them, especially as China's weakness attracted Japanese expansion. Britain and the United States could not cooperate against Japan, encouraging the militarists. The Soviet Union, responding to belligerence in Tokyo, began a great but cautious concentration in Siberia, which blocked Japanese aggression and deflected it south – against Britain. For the first time, the Asia-Pacific system itself split Britain and Japan, if just to a minor extent, rather than unifying them. Japanese competition was a secondary factor in Britain's decision to abandon free trade, which harmed Japan's economy to a tertiary degree.[30] Above all, a revolution in Japanese decision making overturned leaders with whom Whitehall had worked, empowered people who were hostile to it, and turned Japan from being a junior partner in a liberal system to a leading attacker against it. A combination of the weakness of local powers and the strength of economic and political desires, drove Japanese expansion, multiplied

[30] Ishii Osamu, 'Markets and Diplomacy: the Anglo-Japanese Rivalries over Cotton Goods Markets, 1930-36' and John Sharkey, 'Economic Diplomacy in Anglo-Japanese Relations, 1931-41', in Nish and Kibata (eds.), *The History of Anglo-Japanese Relations, Volume 2*, 51-77, 78-111.

by the ability of radical elements in the IJA and the IJN to control policy, and to pursue interests defined by ambition and appetite. Japan became belligerent and erratic, pursuing armed revisionism and militarized mercantilism. It challenged British interests in China and the Asia-Pacific region.

This split in interests and rise in ambitions turned the fractures between Britain and Japan into a break, but the outcome was not preordained. The IJN did not define Britain as a potential enemy until 1936, though it then became one of its primary targets, together with the USN.[31] Only in August 1941, after the United Kingdom, the United States and the Netherlands declared economic sanctions, did Japanese leaders decide on war, to preempt a threat they had provoked. In June 1937, British authorities hoped that relations with Japan might stabilize. With the outbreak of the Second Sino-Japanese War, however, conflict rapidly cascaded; to pass the Yangtse was to cross the Rubicon. The IJA attacked British interests in China, opening an economic war, whilst Whitehall backed the Chinese government to contain Japan. Some Japanese hoped to seize the Asia-Pacific region from Britain, which considered checking its rival through economic sanctions. Britain was forced into humiliations, which caused contempt among foreigners. All Britons could do was take their licks and bide their time, as bilateral and multilateral dangers reinforced each other. Adolf Hitler's Germany tried to drive Japan against Britain. Japan shaped British policy in Europe: thus, in 1937-39, the need to buy time for rearmament drove the Admiralty's support for appeasement. Disasters loomed, as Germany pursued a new alliance with Japan against Britain, and the old allies drifted towards war in the Tienstin crisis. They were avoided only by Japanese inability to reply quickly to Hitler's offer, and his decision to align against Britain with Moscow instead of Tokyo, combined with diplomacy by a few British and Japanese officials.

Britain had to race to recover from self-inflicted weakness, before it could strengthen itself. Its pace was hobbled because key decision-makers like Neville Chamberlain and Treasury officials, were reluctant to act against Japan or Italy, to recognize that states which until recently

---

[31] Yoshio Aizawa, 'The Path Towards an Anti-British Strategy by the Japanese Navy Between the Wars', in Gow, Hirama and Chapman (eds), *The History of Anglo-Japanese Relations, Volume 3*, 139-50; Yamamoto Fumihito, *The Japanese Road to Singapore, Japanese Perceptions of the Singapore Naval Base, 1921-41* (Unpublished PhD dissertation, National University of Singapore, 2009).

had been junior partners were rejecting that status, challenging British rules and the liberal game. Bad relations with them seemed unnatural, rather than the product of power, appetite and maritime decline. This error stemmed from English ethnocentrism, multiplied because Japanese and Italian leaders overestimated their power and imagined they could achieve ambitions which Britons, rightly, thought too out of reach. The Foreign Office and the military services had a better sense of the Japanese problem, because they had to address conflicts every day, and they focused on capabilities rather than intentions.

Yet, in the short period where it had a free hand, Britain did well in rebuilding its sea power against not just Japan, but also Germany and Italy. It completed the Singapore base, a precondition for power against Japan. That base was powerless without a fleet but, given the limits which the London Naval Treaty placed on construction before 1936, Britain could have done little more than it did to restore its maritime power. Between 1936 and 1939, British sea power rose rapidly, because of the fast rebuilding of warships and the capacity to build them. The Admiralty spent in three years all the money which it had been provided for seven, placing five battleships and four aircraft carriers in the water by 1942, to none by the IJN, showing British superiority over Japan in the core of their power relationship.[32] That growth, reinforced by developments in air power, bolstered British confidence against Japan right up to May 1940; Whitehall felt the wind behind its sails. But Britain read its power forward. This success could have mattered only had events worked as the Admiralty hoped; had Britain avoided a world war before 1942, while starting another naval programme. Even then, a hidden weakness loomed. The Admiralty was addressing the problem of naval aviation, but probably could not have overcome it before 1943. Authorities misconstrued these weaknesses, and their significance.

The Japanese problem centred on power. By 1939, Britain was on the way to solving it. By 1940, the collapse of British policy in Europe made that problem worse, yet far behind a German difficulty. Every relationship threw Britain against Japan. Britain could not stop Japan from assaulting its empire in Asia. Japanese leaders grossly underestimated the strength of the United States and overrated that of Germany and Japan. They saw Britain as an enemy and an opportunity,

---

[32] Christopher Bell, *The Royal Navy, Seapower and Strategy Between the Wars* (Houndmills, Basingstoke: Macmillan, 2000).

weak and malevolent. Britain's blockade of Germany, which in effect became one of Japan, intensified by sanctions, created economic war across the Asia-Pacific system. German policy and Russian power, the threats to China and the fall of France, threw Japan at Britain's throat and the latter into the arms of the United States. Britain had to abandon control over its policy in the Asia-Pacific region to Washington. Again, fratricide drove the Anglo-American relationship. President Roosevelt could not form an open alliance with Britain against Japan, leading each side to manipulate the other. In order to show itself alliance worthy in the Pacific, Britain had to convince Washington that it was stronger there than was true, or it thought. In order to shape British policy, Roosevelt had to make Whitehall think he was less likely to help against Japan than he was. Their miscalculations magnified each other – the Anglo-American position in the Pacific was the product of their weaknesses, rather than their strengths. This created a war for mastery of the Asia-Pacific region, which only the United States could win.

Conclusions

Between 1880 and 1941, many states challenged the world system which Britain had created, or their position there. To Britain, Japan ranked small among them until 1937-41, when events in Europe went as wrong as imaginable, and Tokyo acted as badly as possible. The Japanese onslaught of 1941-42 broke Britain's control of the Asia-Pacific region and cut it down below the size of the United States. Yet, Japan merely hastened processes that already were visible, perhaps only by a generation. By 1938, Britain no longer dominated the Asia-Pacific economy, where its ability to shape people and peoples hinged on control over an empire which was wasting away. The Raj, for example, was eroding in India, and might have ended before 1950 even without a war. Japan ensured that the empire passed in humiliation, though during 1944-45 Britain recouped some of that loss, through the greatest investment of power it ever made to the region, with a low return on the investment. Above all, Japan failed to master or to reconfigure the Asia-Pacific system; it sought to redefine it through force but that force, pre-eminently resting on maritime strength, was not meant to support a wider economic system. By 1945, the United States had gained control of the Asia-Pacific and chose to maintain the system, backed by the strength of junior partners, Britain and Japan, once again in a quasi colonial status, though soon becoming one of its dominant economies. China first rejected the system, later rejoined it and, in 2011, challenges its terms of trade and power.

These events, the prehistory of the Asia-Pacific system, illuminate the struggle for power which underlies it today. Compared to 1830-1945, that system is structurally more stable, while the United States possesses as much relative sea power as Britain ever did, but things change. When they do, bilateral relationships move in a multilateral context. Third parties matter, often in unpredictable ways. Intentions cannot always be effected. Unintended consequences cannot be avoided. Successes cause problems. Power is central to systems – marginal changes trigger bigger ones, by attracting friends or causing them to flee. Power always is distributed, but not often balanced; systems resting on a preponderance of power behind the *status quo* are more stable. In a multilateral system, power constantly fluctuates, as do perceptions if it. Periods of transition in power – when a preponderance collapses, or several states race to maximize their strengths, in a pattern of challenge and response – are conducive to crisis and war. Power rises and falls steadily, and also suddenly, driven by crises, decisions, perceptions, ambitions and miscalculations. Tipping points exist. Their location usually is unknown, until the earth moves.

CHAPTER TWO

SEA POWER AND ANGLO-JAPANESE MILITARY RELATIONS, 1863-1923

Haruo Tohmatsu

Old Friends, New Enemies is the title of a classic study on Anglo-Japanese naval relations by renowned historian Arthur J. Marder.[1] The subtitle of the text further read '*Strategic Illusions 1936-1941*'. Indeed, whilst the book focused on the interwar period, this subtitle accurately captures some core features of Anglo-Japanese political and military relations in the period stretching from the Anglo-Satsuma War of 1863 to the termination of the Third Anglo-Japanese Alliance in 1923. This chapter's main aim is to examine how at the regional level, sea power – intended as the ability to project military power and exert political influence across East Asian waters – was key to Anglo-Japanese relations. In the initial years of Japan's modernization, the relationship with Britain started as one between 'client and patron'. It subsequently evolved into a partnership between equals with the overlap of strategic goals in East Asia favouring a military alignment that lasted over two decades.

A second contention of the chapter is that throughout this period, Japan's importance in British strategic calculations grew alongside its ability to exert command of the sea within Asian waters. Initially, a fast-modernizing Japan with the ability to off-set the military presence of other European powers in East Asia offered relief to British forces in the region. Yet, the differences pertaining to Britain's requirements as a global power and Japan's more regionally-focused interests slowly drove the two countries apart. Japanese sea power remained an essential feature in both the making and the demise of the Anglo-Japanese Alliance; it enabled the partnership to develop and, eventually, set the two island nations on their collision course.

---

[1] Arthur J. Marder, *Old Friends, New Enemies: The Royal Navy and the Imperial Japanese Navy. Strategic Illusions 1936-1941* (Oxford: Clarendon Press, 1981).

Britain as a 'Naval Model'

Since the outset of Japanese modernization which began as early as the late 1850s, Great Britain had been seen by many Japanese as a model for Japan. Many Japanese intellectuals and reformers of the *Bakumatsu* period (or 'late Tokugawa Shogunate', 1853-67) were quick to recognize that industrialization and naval power were the foundation upon which Britain had built up national wealth and international prestige. The rapid reconciliation and strong political ties pursued by the clans of *Satsuma* and *Chōshū* with Britain after the military skirmishes of 1863 reflected this recognition.[2] British military presence was also visible and impressive to the Japanese eyes through the stationing of British troops (alongside its French counterpart) in Yokohama for twelve years between 1863 and 1875.[3]

Britain began to occupy a pivotal position in the Japanese political and diplomatic debates after the 1870s when the newly established Meiji Government decided to pursue a new national policy. First of all, it aimed at a revision of the unequal treaties concluded between the Western Powers and the Tokugawa Bakufu in the 1850s.[4] Second, Japanese national security could be guaranteed only if the country were to acquire a degree of naval power.[5] The first aspect required a general modernization (=Westernization) of the political, economic, legal and social systems. In this respect, Britain was regarded as one of the major

---

[2] Earliest British estimate of Japanese military potential was made at the time of wars between Britain and Satsuma and Chōshū clans in 1863. See Tooru Hōya, *Bakumatsu nihon to taigaisensō no kiki: Shimonoseki sensō no butaiura* (Bakumatsu Japan and the crisis of anti-foreign war: The background of the Shimonoseki war, Tokyo: Yoshikawakōbunkan, 2010), 92-160.

[3] For the British military presence in Bakumatsu Japan and Japanese perception of British military and naval force at that time, see Yokohama taigai kankeishi kenkyūkai & Yokohama kaikō shiryōkan (Foreign Relations Study Society of Yokohama & Yokohama Archive of Opening of the Port) ed., *Yokohama eifutsu chūtongun to gaikokujin kyoryūchi* (British and French garrison and foreign compound in Yokohama, Tokyo: Tokyodō shuppan, 1999).

[4] For treaty revision see Turan Kayaoğlu, *Legal Imperialism: Sovereignty and Extraterritoriality in Japan, the Ottoman Empire, and China* (Cambridge: Cambridge University Press, 2010), especially Chapter 3 'Japan's Rapid Rise to Sovereignty', 66-103.

[5] For arguments concerning which service should be the mainstay of national security ('riukushu-kaijū=army first, navy second' versus kaishu-rikujū=navy first, army second) see Taeru Kurono, *Dainippon teikoku no seizon senryaku* (Survival strategy of the Japanese Empire, Tokyo: Kōdansha, 2004), 21-8.

role models alongside Germany and France.[6] Concerning the second aspect, Britain was 'the' model in terms of their similar geography. Both Britain and Japan were mid-sized island states with relatively limited natural resources at home, and that had to turn to maritime trade to increase their national wealth.

Indeed, Meiji leaders well understood that the foundations of British power depended on foreign trade and the navy that protected global commercial routes between the mother country and its dominions and colonies. For that matter, the Imperial Japanese Navy (IJN) was modelled after the Royal Navy (RN), the uncontested dominant global force that ruled the waves, much like the United States Navy (USN) of the present day. From 1873 to 1879, a military assistance mission to restructure the Japanese naval academy led by Commander Archibald Douglas, RN, was particularly instrumental in establishing the foundation of the IJN.[7] Many young and able leaders of the IJN were sent and trained in Britain. Eventually the Japanese Naval Academy in Etajima, in southwest Japan, was built after the architectural model of the naval college in Dartmouth.[8] By the time the Japanese navy was engaged in the First Sino-Japanese War of 1894-95, the country possessed a relatively small but highly competent naval force. Apart from the navy, other main services in banking and financial sector, the postal service, the political parties, and the railways, to mention a few, were developed on the British system. In short, in many ways Great Britain had represented a model for Japanese modernization.

This perception was not particularly reciprocated by Britain taking a strong interest in Japan. It took nearly three decades until Britain began to see the country as a potential ally. The complexities of Japanese domestic politics and institutional reforms of the time suggested caution *vis-à-vis* a close rapprochement with the Asian insular nation. In these three decades, Japan made strenuous efforts to revise the unequal

---

[6] For Japanese images of the British Empire as a role model, see Tetsuya Sakai, 'Hankei to shiteno eiteikoku' [The British Empire as a model] in Yōichi Kibata and Harumi Gotō (eds), *Teikoku no nagai kage* (The Long Shadow of the Empire, Kyoto: Minerva shobō, 2010), 25-48.

[7] Ian Gow, 'The Douglas Mission (1873-1879) and Meiji Naval Education', in J. E. Hoare (ed.), *Britain and Japan. Biographical Portraits, Volume III* (Richmond, Surrey: The Japan Library, 1999), 144-57

[8] For early influence of the Royal Navy on the infant Imperial Japanese Navy see Michio Asakawa, 'Anglo-Japanese Military Relations, 1800-1900', in Ian Gow, Yoichi Hirama and John Chapman (eds), *The History of Anglo-Japanese Relations, 1600-2000, volume 3 The Military Dimension* (Basingstoke: Palgrave Macmillan, 2003), 19-24.

treaties and almost achieved its objective on the eve of the war with China. British recognition of Japan's progress towards modernization was a vital element in the country's success in treaty revisions.[9]

The First Sino-Japanese War

The First Sino-Japanese War of 1894-95 had mixed results for Japan. Rapidly driving Chinese forces out of the Korean peninsula and neutralizing the Chinese fleet, military victory established Japan as a primary regional power in East Asia. In terms of military performance, the British-modelled Japanese fleet overwhelmed the Chinese fleet that exceeded it in both numbers and tonnage.[10] To Japanese institutional authorities, the success proved that they had made the right decision to build their navy after the RN.

By contrast, the primary political and diplomatic goals of the war had not been achieved: the stabilization of the Korean peninsula. The Chinese loss of control over Korea was replaced by increasing influence of imperial Russia. Unexpected Japanese military success provoked undesirable reaction from the European powers that had political and commercial interests in the region. The so-called Triple Intervention, initiated by Russia and joined by France and Germany, forced Japan to renounce to the Lyaodong Peninsula which was to be annexed to the insular country as a spoil of war.[11] Britain did not take part in this political manoeuvre and its lack of initiative to favour Japan's annexation claims was functional to the success of the Triple Intervention. To some extent, Britain had an interest in preserving Japan's power as a significant regional actor. Still, in that occasion it was not sympathetic enough to support Japan. Britain recognized Japan as a rising power in East Asia, but its policy towards the newcomer was not hostile at best.[12]

---

[9] Role of Britain in Japan's struggle for treaty revision, see Chihiro Hosoya and Ian Nish (eds), *The History of Anglo-Japanese Relations, 1600-2000, volume 1 The Political-Diplomatic Dimension 1600-1930* (Basingstoke: Macmillan, 2000), 107-31.

[10] On the conditions of the Chinese navy at the time of the First Sino-Japanese War, see H. P. Willmott, *The Last Century of Sea Power, Volume 1 From Port Arthur to Chanak, 1894-1922* (Bloomington: Indiana University Press, 2009), 19-24.

[11] W. G. Beasley, *Japanese Imperialism 1894-1945* (Oxford: Clarendon Press, 1987), 58-9.

[12] For British attitude towards Japan and the Triple Intervention see Otte, *The China Question*, 312-18.

In 1898, Russia gained formal control of Lyaodong in the form of a
long-term lease from China. Russian presence in the vicinity of Korea
became a permanent source of concern to Japanese interests in Korea, paving the way for Japan to pursue the formation of an alliance with Britain
and, setting the country on a war path with Russia. In the aftermath of the
First Sino-Japanese War, Britain too was perceived to be gradually modi-
fying its approach to Japan. This was explained in two ways. First, Britain
began to realize the threat of Russia towards British regional commercial
interests. As long as Russia remained in Siberia without any large naval
base except for Vladivostok, British strategic advantage in the region was
relatively ascertained. This condition was likely to change with the build-
up of the naval base and related military installations in Port Arthur.

Losing the Lyaodong peninsula and the unexpected increase of Russian
influence in Korea convinced the Japanese leadership of the necessity
to develop stronger political ties with an influential power as a way to
increase its international profile and secure the country's diplomatic
aims.[13] In this context, many senior Japanese leaders still had the vivid
memory of what had happened in 1861. In the summer of that year when
the Russian fleet made an attempt to occupy the islands of Tsushima, a
RN warship was dispatched to the spot and the Russians were immediately forced to leave the islands. To the Japanese, it was obvious that the
territorial ambitions of Russia, and to some extent of France, towards
Bakumatsu Japan were kept in checked by British political and military
pressure.[14] Britain did have commercial interests regarding Japan but it
did not have any primary appetite for territorial expansion. Britain was
a global maritime Empire and by the 1860s, it was consolidating its vast
commercial web, rather than expanding its possessions.[15]

Japan had learned an important lesson from the war with China.
Its leadership understood the limits of an emerging regional power in
the international politics of great power imperial competition in Asia.
Unless it succeeded in linking its own interests to that of an influential
state actor, it would have been difficult to secure the goals of its diplomatic agenda in the region.

---

[13] Japanese leaders seriously learned this lesson of the First Sino-Japanese War. See
Kurono, *Dainippon teikoku no seizon senryaku*, 56-62.

[14] For Anglo-Russian relations regarding Japan in the 1860s see Hamish Ion, 'Days of
Seclusion' in Hosoya and Nish eds, *The History of Anglo-Japanese Relations, 1600-2000,
volume 1*, 4-7.

[15] As British efforts were concentrated in stabilizing India after the Great Mutiny,
there was not much scope for conquering new territories.

The First Anglo-Japanese Alliance

In the literature, it is well-known that a shared perception of the Russian threat in East Asia was a primary driver in the formation of the Anglo-Japanese Alliance.[16] An additional reason that motivated Britain to conclude, after a century of 'splendid isolation', a formal treaty with Japan concerned its slowly eroding military and naval dominance in the region.

Britain's imperial network extended on a global scale, but at the turn of the twentieth century its commitments in Europe, Africa and India were increasing. Two successive colonial wars in South Africa and north China had revealed the limits of an overstretched global power. As its army was primarily deployed in the Boer War since 1899, Britain was seriously short of ground forces when the Boxer Rebellion broke out in north China in the following year. Approximately 6,000 men were the maximum contribution Britain could muster and dispatch in that theatre.[17] The mainstay of the multi-national coalition army that eventually suppressed the Boxers was composed of Russians and Japanese forces, about 22,000 and 12,374 strong, respectively.

In addition to the strains imposed by Britain's colonial commitment on the army, increasing tensions between the continental powers of Europe and Britain at home meant less flexibility for the RN to deploy major detachments of surface forces in Asian waters. British naval superiority was still undisputed: Britain possessed 1.52 million naval tonnage in 1902 against French 0.6, Russian 0.39, German 0.36, American 0.3, and Japanese 0.24.[18] However, insofar as the number of

---

[16] For the formation, execution, development of the Anglo-Japanese Alliance in general, see Ian Nish, *The-Anglo-Japanese Alliance: The Diplomacy of Two Island Empires, 1894-1907* (London: Athlone Press, 1966); Phillips P. O'Brien ed., *The Anglo-Japanese Alliance, 1902-1922* (London: Routledge Curzon, 2004); Yōichi Hirama, *Nichiei dōmei* ('The Anglo-Japanese Alliance, Tokyo: PHP, 2002). For British evaluation of Japanese military and naval capability, see Philip Towle, 'British Estimates of Japanese Military Power 1900-1914', in Philippe Towle ed., *Estimating Foreign Military Power* (London: Croom Helm, 1982). John Ferris, 'Double-Edged Estimates: Japan in the Eyes of the British Army and the Royal Air Force, 1900-1939', in Gow, Hirama and Chapman eds, *The History of Anglo-Japanese Relations, 1600-2000, volume 3*, 91-108.

[17] On conditions of the British army at the time of the Boer War and Boxer Rebellion see Edward Spiers, 'The Late Victorian Army', in David Chandler ed., *The Oxford Illustrated History of the British Army* (Oxford: Oxford University Press, 1994), 189-214.

[18] On British naval strength and strategy at the turn of the century see Paul Kennedy, *The Rise and Fall of British Naval Mastery*, Third edition (London: Fontana Press, 1991), 243-82

capital ships (battleships and cruisers) in East Asia was concerned, in April 1901 Britain had only four against Russian eleven and French two. The overall balance was one in which Britain fielded four main surface vessels against a potential Franco-Russian fleet of thirteen.[19] For Britain, it was becoming evident that there was the need to find a means to fill this capability gap.

By that time, Japan possessed a fleet of nine capital ships. This number was slightly inferior to the combined Franco-Russian force in the region; yet, its forces were sufficient to maintain Franco-Russian activities at bay. The IJN was a highly professional force with proven tactical potential and battle-hardened manpower. If British naval power were to be joined by that of Japan, the number of capital ships of a combined Anglo-Japanese coalition would make fifteen against Franco-Russian thirteen.[20] In addition to the capital ship numbers, an Anglo-Japanese coalition would have an additional geo-strategic advantage. Britain and Japan had two large naval bases in Weihai and Sasebo while the Franco-Russian alliance could count only on one in Port Arthur. French bases in Indo-China were too distant to make its navy operate effectively and protractedly in the East China and Yellow Seas.

If the British rationale for approaching Japan at the turn of the century was essentially a strategic one, Japan had additional reasons to seek an alliance with Britain. First, Japan as an infant industrializing nation needed technological assistance from one of the world's most advanced industrial states, Great Britain. Second, Japan was in great need of wider financial support in order to fight a possible, if not probable, war against Russia (and, if necessary, France) in the region. Third, an equal partnership with Britain in the form of an alliance treaty would greatly elevate Japan's status in the international arena. This last fact was regarded by Japanese authorities with particular interest as it was instrumental to complete the revision of the unequal treaties concluded by the *Bakufu* in the closing days of the Tokugawa regime.[21]

---

[19] For British naval weakness in the East Asia see Willmott, *The Last Century of Sea Power*, Chapter 4, 53-61.

[20] For British naval strategy prior to the Russo-Japanese War, see Arthur Marder, *The Anatomy of British Sea Power: A History of British Naval Policy in the Pre-Dreadnought Era*, 1880-1905 (New York: Octagon Press, 1976). For British Far Eastern strategy concerning Russia, see Keith Neilson, *Britain and the Last Tsar: British Policy and Russia, 1894-1917* (Oxford: Oxford University Press, 1995), especially Chapter 7.

[21] About psychological impact of the Anglo-Japanese Alliance see Ian Nish, 'Echoes of Alliance 1920-30', in Hosoya and Nish eds, *The History of Anglo-Japanese Relations, 1600-2000, volume 1*, op. cit., pp. 255-78.

The Russo-Japanese War was fought under the aegis of the First Anglo-Japanese Alliance and the process and outcome of the war proved the accuracy of the strategic judgment made by both Britain and Japan. Britain succeeded in limiting Russian southern advance without dispatching any ground forces or components of the fleet to East Asia. Britain did offer support to Japan within the existing legal framework, obstructing the sale of Argentine warships to Russia, and sought to obstruct the preparations of the Russian Baltic Fleet at Dogger Bank at the outset of its doomed expedition to Tsushima.[22]

On the other hand, the Anglo-Japanese Alliance did not seriously jeopardize Anglo-Russian relations. Recent scholarship proved that Britain made efforts to maintain relatively stable, if not amicable, relations with Russia and succeeded.[23] The war itself had the additional effect of making France reconsider to a certain degree its long-standing colonial rivalry with Britain. The unexpected Russian unimpressive military and naval performance against the Japanese forces in the opening months of the war had much to do with France's closer cooperation with Britain, paving the way to the Entente Cordiale in the spring of 1904.[24]

The Second Anglo-Japanese Alliance

As the beneficial effects of the alliance became evident, negotiations for its renewal started as early as February 1905.[25] By that time, the Imperial Japanese Army (IJA) had captured Port Arthur and the IJN had firmly secured the sea lanes of communication (SLOCs) between

---

[22] For British assistance to Japan for purchasing cruisers from Argentine, see Hiraku Yabuki, 'Britain and the Resale of Argentine Cruisers to Japan before the Russo-Japanese War,' War in History, Vol. 16, 2009:4, 425-46. Also for British naval assistance to Japan during the Russo-Japanese War, see Philippe Towle, 'British Assistance to the Japanese Navy during the Russo-Japanese War of 1904-5,' Great Circle II (1980), 44-5.

[23] Robert K. Massie, Dreadnought: Britain, Germany, and the Coming of the Great War (New York: Ballantine Books, 1991), 595-6

[24] Hosoya and Nish eds, The History of Anglo-Japanese Relations, 1600-2000, volume 1, 173, 205. The German factor was also important in the conclusion of the Entente Cordiale. For that matter, for German involvement in the Russo-Japanese War, see John Steinberg, 'Germany and the Russo-Japanese War,' American Historical Review LXXV (1970), 1965-86.

[25] Keith Neilson, '"A Dangerous Game of American Poker": The Russo-Japanese War and British Policy,' Journal of Strategic Studies XII (1989), 63-87; B. J. C. McKercher, 'Diplomatic Equipoise: The Lansdowne Foreign Office, the Russo-Japanese War of 1904-1905, and the Global Balance of Power,' Canadian Journal of History XXIV (1989), 299-339.

the Japanese homeland and Korea/Manchuria. The final result of the war was still unforeseeable, yet the possibility of a Russian victory was less likely.

Britain now had strong confidence in Japanese military and naval capabilities; in that respect it was a reliable ally. This fact explains the drastic change of nature of the Second Alliance which came into being in August 1905. First, the revised Alliance was more mutual in nature. Both parties had immediate obligations to aid each other in case of the break out of hostilities with a third party. The primary potential common threat remained Russia, though neither country had reason to exclude Germany, Turkey, or even the United States as potential regional competitors in East Asia.[26]

Britain's main strategic concern was, in addition to the Anglo-German naval rivalry, the defence of India. India, especially its north-western border was vulnerable to attacks by Russia via Afghanistan. Local chiefs of Afghanistan were not supportive of British authority and some even showed favour to the Russians. Repeated British attempts to stabilize this area since the mid-nineteenth century had always failed. Afghanistan and the north-western border corresponded to a British 'interest line' to protect India, the 'sovereignty line' of the Empire. In this context Britain had a high expectation of the IJA's assistance in case of a war with Russia in this region. Battle-hardened Japanese troops would be a powerful and reliable ally in repelling a Russo-Afghan onslaught to India.[27]

Japan, on the other hand, had equally major expectations in British aid in case of another war against Russia in the future. Considering the logistical difficulties due to the vast oceanic distances between the two island empires, it was quite doubtful as to whether the Japanese would be able to dispatch a substantial number of ground forces to India, and British reciprocating in the same way by sending expeditionary forces to Manchuria.[28] The true significance of the revised Alliance was

---

[26] US attitude towards the Second Alliance was naturally not favourable. See Nish, *The Anglo-Japanese Alliance*, 328-30.

[27] For British expectation on Japan regarding the defence of India see Nish, *The Anglo-Japanese Alliance*, 314-20.

[28] Concerning the India factor in the revision of the Anglo-Japanese Alliance, during the Russo-Japanese War, a Japanese military attaché in India was gathering detailed information of the British military capacity in the country. See Michio Yoshimura, 'Nichirosensōki niokeru eiryō indo rikugun chūzaibukan hōkoku' [Reports by a Japanese Army Attaché Posted in India during Russo-Japanese War], *Gaikōshiryōkanpō* [Journal of the Diplomatic Record Office] Vol. 18, September 2004, 25-37.

perhaps in the fact that a closer cooperation between the two countries would deter Russia from making dangerous gambles on two different fronts. In this sense, the Second Anglo-Japanese Alliance was a strategy of containment implemented by two maritime powers against a continental power.

After 1907, with Russia and Japan signing an entente treaty, the possibility of a Russian war of revenge against Japan in Manchuria in the near future declined. In the same year, Britain and Russia reached a similar agreement and the latter's attack on India became unimaginable. The Anglo-French Entente Cordial of 1904 also functioned to bring Britain and Russia to the same camp. The French diplomatic approach to Japan and the conclusion of a Franco-Japanese entente in 1907 further reduced the possibility of another Russo-Japanese war. Financial dependence on France in the aftermath of a consuming conflict, made it difficult for Russia to venture into another conflict against Japan without French and British approval. It is ironic that the fierce confrontation in 1904-05 between Russia and Japan brought a transformation in international politics that eventually changed former foes to allies.[29] The Second Anglo-Japanese Alliance thus rapidly lost its original rationale in the ensuing years.

The Third Anglo-Japanese Alliance

The sequence of events unfolding throughout the first half of the 1910s represented a turning point in the Anglo-Japanese military relations. In August 1910, Japan formally declared the annexation of Korea. This was executed under the full recognition of other great powers including Britain and the United States.[30] By putting Korea firmly under its direct control, Japan achieved the long-sought national goal of securing the 'sovereignty line', or *shuken-sen*, by holding a strong position on the Korean 'interest line', or *rieki-sen*, as Prime Minister Yamagata Aritomo stated in the Diet in 1890.[31] In 1911, Britain and Japan concluded a new trade and navigation treaty which was followed by similar

---

[29] Two voluminous works on the Russo-Japanese War study recently examined this theme. See John Steinberg, David Wolff, et al. eds, *The Russo-Japanese War in Global Perspective: World War Zero*, volumes 1 & 2 (Leiden: Brill, 2005 & 2007).

[30] Peter Duus, *The Abacus and the Sword: The Japanese Penetration of Korea, 1895-1910* (Berkley: University of California Press, 1998).

[31] For well-known terms of 'high imperialism geopolitics' by Yamagata see W. G. Beasley, *Japanese Imperialism 1894-1945* (Oxford: Clarendon Press, 1987), 45-6.

agreements stipulated with other powers, completing Japan's long-term effort of revising the unequal treaties.[32] The year 1911 saw a change of monarch in Great Britain. The short reign of Edward VII drew to an end, and his son, King George V, succeeded him. Japan was also to end the turbulent Meiji era and see the coronation of a new sovereign within a year, the Emperor Taishō.

Domestic changes were coupled with the emergence of a new regional power in East Asia. The post-Russo-Japanese War years witnessed the rapid ascendance to power of the United States in international politics. After acquiring the Philippines in 1898 from a declining Spanish Empire and participating in suppressing the Boxer Rebellion in 1900, the US emerged as one of the major actors in the Asia-Pacific scene. American mediation of the two belligerents at Portsmouth in September 1905 eventually brought an end to the Russo-Japanese War and this further contributed to establish American prestige. Receding Russia and rising America were two prominent features that contributed to reshape the East Asian geopolitical context throughout the duration of the Second Anglo-Japanese Alliance. Under such circumstances, the Anglo-Japanese Alliance was revised for the second time.

The initiative came from Britain. Although there was very little possibility of an Anglo-American confrontation of serious nature, the second revision did not in theory exclude the US from the list of potential opponents. For Britain, antagonism with the United States was to be avoided. Provided that Washington was against the idea of 'alliance diplomacy based on balance-of-power perception of the world', and that the Anglo-Japanese naval cooperation surpassed American naval power in the Pacific, Britain carefully had the United States removed from the aforementioned list.

This was not particularly satisfactory to Japan, as US-Japan relations had been less than cordial after the Russo-Japanese War. The two rising powers at the opposite ends of the Pacific were now looking at each other as potential rivals – an idea that gauged the views of Japanese naval planners. The visit of the US 'White Fleet' to Japan in 1906 was received in Japan as a demonstration of might rather than of goodwill. In fact, both the USN and IJN began to envisage plans against each other shortly after the Russo-Japanese War. The two navies envisaged a Tsushima-type

---

[32] Peter Lowe, *Great Britain and Japan, 1911-1915: A Study of British Far Eastern Policy* (London: Macmillan St. Martin's Press, 1969), 21.

grand battle in the western Pacific.[33] The exclusion of the United States from the list of potential threats meant that Japan could no longer expect British assistance in case of the advance of an American armada in the western Pacific. This was perhaps one the elements that contributed to the development of an 'inferiority paranoia' of the IJN towards the USN. Indeed, in the following years, this 'hypothetical' enemy became the paramount preoccupation of Japanese naval planners, with an 'interception strategy' and a fanatically rigorous training becoming the hallmark of the service's attempt to meet the American challenge.[34]

The overall evaluation of the Third Anglo-Japanese Alliance is difficult. In military and naval terms, the Alliance had lost part of its original momentum: Russia and France were no longer opponents to Britain and Japan. The US was excluded from the object of the treaty, and Japanese apprehension *vis-à-vis* the emerging Pacific power was planting the seeds for the need to strengthen Japan's position in the Pacific. Within the British Empire, there was scepticism regarding the Anglo-Japanese Alliance. The southern Dominions, especially Australia, were anxious about Japanese expansion and had ambivalent attitudes towards the extension of the Alliance. On the one hand, it protected the Dominions from Japanese aggression, but on the other, it weakened the British position regarding Japan's stance towards them. Taking all these new elements into account, the spectre of a German threat in the Pacific seemed the only real concern for all parties involved to continue the Alliance.[35]

The Anglo-Japanese Alliance and the First World War

By the time of the outbreak of the Great War in the summer of 1914, in the eyes of many the Anglo-Japanese Alliance was something of a 'hollow union'. In reality, Russia was now on the side of the Entente Powers

---

[33] For interwar US-Japan naval strategic planning see Edward S. Miller, *War Plan Orange: The U.S. Strategy to Defeat Japan 1897-1945* (Annapolis, MD: Naval Institute Press, 1991).

[34] David Evans & Mark Peattie, *Kaigun: Strategy, Tactics, and Technology in the Imperial Japanese Navy 1887-1941* (Annapolis, MD: Naval Institute Press, 1997).

[35] Ian Nish, *Alliance in Decline: A Study in Anglo-Japanese Relations 1908-1923* (London: Athlone Press, 1972). For British strategic calculations from the conclusion of the Anglo-Japanese Alliance until the breakout of the Great War, see Keith Neilson, 'The Anglo-Japanese Alliance, and British Strategic Foreign Policy, 1902-1914,' in Philippe P. O'Brien ed., *The Anglo-Japanese Alliance, 1902-1922*, 50-3. For British intention and assessment of the Third Anglo-Japanese Alliance, see Lowe, *Great Britain and Japan, 1911-1915*.

with which Japan had political and military collaborations. The German presence in Asia and the Pacific was not a sufficiently serious threat preventing Britain to pursue its war aims in Europe and the Middle East.

As such, at the outset of the hostilities in Europe, Britain's main concerns in relation to Japan were twofold. It wished to limit the Japanese military participation to the conflict to a degree that would not erode British interests in East Asia. Similarly, it sought to avoid an increase in the Japanese threat to the Pacific Dominions, in particular to Australia.[36] Conversely, Japan envisaged in the war a chance to consolidate its regional leadership and took advantage of the Anglo-Japanese Alliance to enter the war on the side of the Entente Powers. As the expression *Taishō no ten'yū* (Grace of Heaven of the Taishō period) by leading politician Katō Takaaki well summarized, a devastating war in Europe was a great opportunity. Japan could expand its sphere of influence and increase its wealth in East Asia and the Pacific.[37] On 23 August 1914, Japan declared war on Germany and Japanese forces swiftly moved toward German bases in the Shandon peninsula and Micronesia.[38]

Military operations in Micronesia were a one-sided show by the IJN, as modest German garrisons scattered across the small islands surrendered to IJN landing forces without substantial resistance. By contrast, operations in Shandon proved to be more demanding for the IJA expeditionary forces facing a much more organized resistance by the German garrison stationed in the fortress of Qingdao.[39] In relation to Anglo-Japanese relations, the capture of Qingdao represented one case of wartime cooperation in East Asia as a small unit of the British Indian Army fought alongside the IJA units. British participation in the Shandon operation had more of a political significance than substantial military contribution. However, the issue of command structure brought about a degree of friction between IJA and British military leaders.[40] This example raises the question of what would have been the

---

[36] John Beaumont (ed.), *Australia in War 1914-1918* (St Leonards, NSW: Allen & Unwin, 1995), 1-7.

[37] Frederick R. Dickinson, *War and National Reinvention: Japan in the Great War, 1914-1919* (Cambridge, MA: Harvard University Asia Center, 1999).

[38] For Japanese strategy in the Great War see Yōichi Hirama, *Dai'ichiji sekaitaisen to Nippon kaigun* (The First World War and the Japanese Navy, Tokyo: Keiō Gijuku University Press, 1998).

[39] Spencer C. Tucker, *The Great War* (Bloomington: Indiana University Press, 1998), 195-7.

[40] For British participation in operation in Shandon see Hirama, *Dai'ichiji sekaitaisen to Nippon kaigun*, 157-66.

impact on the alliance of more regular and larger scale cooperation had Britain and Japan sent ground forces to Manchuria and India respectively under the obligation of the Second Anglo-Japanese Alliance.

As the conflict grew into a prolonged war of attrition, Britain seriously suffered from shortage of forces. Even after mobilizing men and materiel from the Dominions and India, Britain still had the need of additional reinforcement. As late as the spring of 1917, the United States and Japan were the only two major powers that had been left almost untouched by the ravage of war. The US had entered the war in April 1917. However, it was not fully prepared to take part in military actions and it was most likely that another year or so would be necessary to turn its potential power to actual.[41] Japan was the only power from which Britain could expect immediate and substantial military support. In this context, it is worth reviewing the British manoeuvres to persuade Japan for an increased commitment to the war, as reported by Arthur Balfour.[42]

Balfour initially worked in the Committee of Imperial Defence, and subsequently as First Lord of the Admiralty, during the premiership of Prime Minister Herbert H. Asquith. With the formation of a war cabinet under David Lloyd George, Balfour was appointed Foreign Secretary. Alongside Lloyd George and Jan Christian Smuts, Balfour showed powerful leadership in running the wartime British Empire. Whilst Lloyd George chiefly dealt with home and European affairs, Balfour oversaw other matters, including the negotiations to gain naval and military support from a reluctant Japan.

In December 1916, Balfour was appointed as Foreign Secretary of the Asquith Cabinet and began to sound out the possibility of gaining naval assistance from Japan. The initial idea was to purchase two of the newer battle cruisers from the IJN. Knowing the IJN's reluctance to sell its assets, the purchase plan was changed to a time-fixed lease. Even in this case, the Japanese Government rejected the British request on the ground that capital ships were indispensable for the defence

---

[41] Tucker, *The Great War*, op. cit., pp. 131-6.

[42] On British request of military assistance to Japan in general and on Balfour's role on that matter in particular, this paper owes much to Takeshi Sugawara, 'Balfour and the Anglo-Japanese Alliance during the Great War' (Unpublished M.A. thesis, University of East Anglia, 2008). On Japanese reaction to British requests for military assistance, see Akio Nagai, 'Dai ichiji sekaitaisen niokeru ōshu sensen hahei yōkyū to nihon no taiō' (Japanese Response to the Allied Demand for the Dispatch of Troops during World War I), *Senshibu Nenpō*, 1998:1, National Institute for Defense Studies, 9-21. V. H. Rothwell, 'The British Government and Japanese Military Assistance 1914-1918', *History*, Vol. LVI, 1971:186, 34-5.

of Japanese home waters. Japan's reluctance to engage outside the Asian theatre was not well-received by the Entente Powers, especially considering the country's modest contribution and the actual economic profit of the Japanese economy from the war. 'Showing the flag' by Japan was much expected. Capital ships that embodied the might of a nation had a particularly symbolic meaning. They were visible to the general public more than small auxiliary ships such as destroyers and escorts. Japan's refusal to sell or lease battle cruisers to Britain harmed its national image among the British.

Balfour's requests to Japan changed again, and military assistance was sought by means of dispatching IJA divisions to the Eastern Front. By the end of 1916, the Russian Army was at the verge of collapse, suffering from the repeated blows inflicted by German armed forces. Balfour played the role of facilitating negotiations between the Russian and the Japanese governments regarding the dispatch of IJA troops to the Eastern Front via Siberia. Russia was already provided with weapons by Japan and hoped to limit Japanese assistance to that degree.[43] On the other hand, Russia was desperately short of competent forces and eventually agreed to come into negotiations with Japan through British channels. Russo-Japanese negotiations however soon came to a deadlock for two reasons. First, Japanese requests in exchange of military assistance were to Russian eyes unacceptable. When Russia showed its willingness to cede the northern half of the island of Sakhalin, still in Russian hands after the Portsmouth settlement of 1905, Japan asked for the control of the railways in Manchuria up to Harbin and for the dismantling of Russian naval installations in Vladivostok. Second, the deteriorating situation in Russia after the revolution in February 1917 seriously jeopardized the railway networks in the country. By spring 1917, transporting substantial numbers of IJA units to the Eastern Front via the Trans Siberian Railway was not feasible. When the plan of sending IJA to sustain the Eastern Front failed, Balfour considered an alternative deployment of IJA troops, this time to Mesopotamia.

By March 1917, British forces had captured Baghdad and the Arab revolt supported by Britain was pinning down a great number of Turkish troops in Arabia. The possibility of a powerful Turkish counter offensive to retake Baghdad was still expected. Balfour suggested the

---

[43] Concerning Anglo-Russian relations during the Great War and the position of the Anglo-Japanese Alliance in this context, see Keith Neilson, *Strategy and Supply–The Anglo-Russian Alliance 1914-1917* (London: Allen & Unwin, 1984).

deployment of IJA divisions in Mesopotamia against the Turks. India was an important factor in this scenario. Should IJA troops be sent to Mesopotamia, India would have represented the rear base from which Japanese were to operate. Means of transportation to India and the logistical capacity of the country were key issues to implement the plan. After careful investigation of these issues the conclusion of the British Government was negative. In addition to technical and administrative difficulties, the India Office and the Indian Governor General were against the plan for political reasons. Although Japan had fought a war with Russia under the concept of 'Civilized Japan against despotic Russia', Japan's victory was perceived in many parts of the non-European world as a victory of a coloured race over a white Christian race. The fear was that the arrival and stationing of a large number of Japanese troops in India would favour local Indian nationalist initiatives. This consideration was not ill-conceived since in 1916, Britain had suppressed a mutiny of Indian soldiers in Singapore.[44]

British post-war designs for the Middle East acted as an additional limitation to any plan to dispatch the IJA to Mesopotamia. In early 1917, a secret agreement was concluded between Britain, France and Russia. The basic goal was the partition of the Ottoman Empire and the establishment of an international control of Jerusalem.[45] Had the IJA taken part in operations in Mesopotamia against the Turks, Japan would certainly have had a voice in the shaping of a post-war Middle East. There could have been a Japanese mandate in Mesopotamia under the aegis of the League of Nations.[46] As in the case of India, the Japanese presence in the Middle East was regarded as a potential destabilizing factor, stirring Arab nationalism in the region. This was of course an undesirable situation for Britain. The idea of dispatching IJA divisions to Mesopotamia was therefore abandoned.

The only remaining choice was to send IJA troops to Vladivostok to suppress Bolsheviks after they took power in Russia after October 1917.

---

[44] Byron Farwell, *Armies of the Raj: From the Great Indian Mutiny to Independence 1858-1947* (New York: W. W. Norton, 1989), 236-47.

[45] A. L. Macfie, *The End of the Ottoman Empire 1908-1923* (London: Longman, 1998), 161-72.

[46] Although geographically remote, in 1920 the US was offered a mandate for Armenia by the League of Nations but they declined it. For the American mandate in Armenia, see James B. Gidney, *Mandate for Armenia* (Kent, Ohio: Kent State University Press, 1967). However, this case suggested that Japan could have been given a mandate in the Middle East if it had had contributed substantially to the Entente Powers' war against the Ottoman Empire in Mesopotamia.

Failed British attempts to bring IJA to the Eastern Front or Mesopotamia unexpectedly paved the way to the Siberian Intervention.[47]

Conclusions

A fundamental change in the post-war Asia Pacific region occurred with the agreements signed at the Washington Conference of 1921-22. As a result of this conference, the Third Anglo-Japanese Alliance came to end and the new Four Power Treaty was concluded between Britain, the US, Japan and France. The document had replaced a military alliance with a *de facto* statement of intent to maintain the *status quo* in the Pacific region. It was widely known at that time that the new treaty received strong support from the United States, aiming at the dissolution of the Anglo-Japanese Alliance.[48] There was a certain sense of disenchantment in Britain and Japan at the abolishment of the Alliance. Britain was realizing the decline of its power, whilst Japan sensed the start of the new age of Anglo-American domination.

In Washington, alongside the new Four Power Treaty, the Naval Limitation Treaty between the five leading naval powers was arranged (Britain, the United States, Japan, France and Italy). Article 19 of the Naval Treaty stated the freezing of military and naval installations of the signatories' island possessions in the Pacific, though Singapore base was carefully excluded from it. It was a British attempt to reorganize its Asia-Pacific strategy with a power base in Singapore. Dispatch of a powerful fleet to Singapore from home waters in the case of crisis in the region would protect British interests in East Asia, and shield its southern Dominions from a potential Japanese threat. On paper, the Singapore strategy seemed assuring but in reality it turned out to be not feasible.[49]

---

[47] For Anglo-Japanese relations during the Siberian Intervention, see Hiroaki Katō, *Gureebusu Shireikan no Mita Shiberia Shuppei: Han Borishevikiseiken Womeguru Kokusaikankei* (Siberian Intervention through the Eyes of General Graves: International Relations over Anti-Bolshevik Government – Unpublished Master's thesis, Sophia University, Tokyo 2008), 27-9. Cooperation between British and Japanese army units on theatre was generally amicable. By contrast, Japanese had friction with the Canadian contingents of the British Commonwealth force. In this respect, the Siberian Intervention revealed different perceptions within the British Commonwealth towards the Anglo-Japanese collaboration.

[48] Erik Goldstein and John Maurer (eds.), *The Washington Conference, 1921-22: Naval Rivalry, East Asian Stability and the Road to Pearl Harbor* (London: Frank Cass, 1994), 286-8.

[49] Goldstein and Maurer, *The Washington Conference*, 67. On the Singapore strategy and Australia see John McCarthy, 'Singapore and Australian Defence 1921-42', *Australian Outlook*, Vol. 25, 1971:2, 165-80.

The relations between Britain and Japan began with short but intense military clashes between British forces and the clans of Satsuma and Chōshū in 1863. During Japan's modernization after the 1870s, Anglo-Japanese relations were distinguished by a patron-client relationship. British recognition of Japanese military capability and discipline in the First Sino-Japanese War and Boxer Rebellion gradually transformed the relation of the two nations into a senior-junior partnership. This was particularly obvious in the conclusion of the Anglo-Japanese Alliance. This alliance began as collaboration between global and regional powers. Common strategic aims united the two island empires: checking Russian expansion in East Asia; and containing German penetration into China and the Pacific. The strategic effectiveness of the Alliance was heavily dependent on naval power. Japanese successes at Port Arthur, Yellow Sea and Tsushima proved the effectiveness of the British naval model. This in turn, empowered Japan with the political capital and strategic mobility to operate across the region as a major military actor. The ability of the Japanese navy to project the country's military power in the region stood at the heart of its value as a partner to Britain.

Sea power was a vital pillar of the alliance also for an additional reason. Both Britain and Japan maintained relatively small ground forces, and mutual support in the case of a Russian invasion of Manchuria or a Russian attack on the north-western front of India was regarded as of great strategic importance. Mobility to redeploy forces across the Asia Pacific and sustain them logistically was an equally important task. A task made difficult as a political divide drove the two countries apart and that became apparent during the Great War. Japan's naval assistance and limited ground support advanced Japanese interests in East Asia, secured the country new possessions and weakened the British and, by the end of the war, mutual perceptions and strategic priorities had changed. The emergence of the US and its advocacy for a 'New Diplomacy' in the aftermath of the Great War eventually led the Alliance to its end. The termination of the Alliance in 1923 opened the curtain on a new chapter of the military relations. A 'rivalry' in East Asia was to emerge between Britain and Japan. The strategic reach and mobility that had made naval power so crucial to the alliance was to become a key factor in reciprocal threat perceptions.

CHAPTER THREE

BRITAIN'S STRATEGIC VIEW OF JAPANESE NAVAL POWER, 1923-1942[1]

Douglas Ford

The collapse of Britain's Asiatic bastion at Singapore in February 1942 has been described by a prominent historian as 'the greatest national humiliation suffered by Britain since (Cornwallis' defeat by the American revolutionary forces at) Yorktown', over a century-and-a-half earlier.[2] For British politicians and defence planners, the defeat signified the demise of their seemingly impregnable position as the dominant power in East Asian waters. Perhaps most importantly, the loss of Singapore was the first link in a chain of events that culminated with the dismantling of Britain's global empire. The episode has therefore been the subject of numerous studies by scholars of British defence policy, as well as specialists of Anglo-Japanese relations prior to the Pacific War.

This chapter explores the reasons for the setbacks which Britain suffered during the opening phases of the conflict, examining the political and military factors which dictated British strategy against Japan during the approximately twenty years intervening between the termination of the alliance in 1922 and the outbreak of hostilities. The chapter will first examine how the rise of Japanese naval power affected Britain's strategic situation. It will subsequently explain the rise of the so-called 'Singapore strategy' during the early 1920s, and the RN's attempts to devise a viable plan for protecting Britain's imperial interests. The main focus will be on the reasons why the British failed to develop an adequate course of action for containing the rising tide of Japanese expansion. The shortfall was not caused by incompetence on the part of British leaders, but more a result of the fact that Britain's resources were overstretched, to the point where it could not protect its territories.

---

[1] Large sections of the chapter have already appeared in Douglas Ford, *Britain's Secret War against Japan, 1937-1945* (Abingdon, OX: Routledge, 2006).
[2] S. Woodburn Kirby, *Singapore: the Chain of Disaster* (London: Cassell, 1971), xiii.

An additional factor which prevented Britain from formulating an effective defence plan was a poor knowledge of Japan's military capabilities and strategy. The available intelligence appeared to suggest that Tokyo was reluctant to risk a war in which it might have to face the combined strength of the US and the British Empire, since Japan's armed forces did not have the capacity to prevail in such a conflict. Consequently, the defence establishment developed a complacent attitude, and pursued a strategy which hinged on the expectation that, if the western powers took a firm stand, the Japanese would refrain from invading Britain's territories. Furthermore, even if the Japanese did choose to attack, Britain was unlikely to face serious problems in safeguarding its empire. While faulty intelligence was not the root cause for Britain's setbacks, it led the British to mistakenly believe that their strategy against Japan was sound.

The Genesis of the 'Singapore Strategy'

For Britain, East Asia formed an integral part of its empire for a host of reasons. The first of these was the colonies of Malaya and Borneo, which provided essential raw materials for British industries. In addition, China provided lucrative commercial and trading opportunities, while from the strategic point of view, Southeast Asia commanded the lifeline to the Pacific Dominions of Australia and New Zealand. Given the economic and strategic importance of the Asia-Pacific region, maintaining a secure lifeline to the area was essential for Britain's survival as a Great Power. Yet, until the early part of the twentieth century, the British did not need to worry themselves about maritime threats against their Far Eastern empire, mainly because the RN reigned supreme. Sea power was key to British economic and military dominance. Britain's only rivals were on the European continent, and none of them possessed a fleet strong enough to challenge its superiority. The RN also commanded the exit points from Europe to the outside world, namely, the Atlantic approaches, North Sea, Straits of Gibraltar and Suez Canal. In the event of a war against any of Britain's imperial rivals, the navy could easily curtail the movement of enemy forces abroad.

By the 1900s, with the rapid emergence of Japan as a Great Power and the emergence of the IJN, Britain's position faced a challenge that could not be ignored. For the first two decades of the twentieth century, Anglo-Japanese rivalry was alleviated by the alliance which the two powers had signed in 1902. Under the terms, both powers were to refrain from encroachments on the other party's interests, while remaining neutral in the event of a conflict against a third power.[3] The alliance laid the basis for cordial relations, and during the First World War, Britain's strategic position was significantly eased when Japan agreed to fight on the Allied side, and took on the task of neutralizing German forces in Asia and the Pacific. The contribution which Japan made to the Allied war effort, however, was quickly offset in May 1915, when the Foreign Ministry issued the Twenty-one Demands, which insisted on a special position in China and privileged access to its trade markets. For the remainder of the conflict, Tokyo continued to make similar demands, raising doubts in London as to whether Japan could be counted upon to respect the integrity of British interests in the Far East.[4]

It was amidst this atmosphere of deteriorating Anglo-Japanese relations that Singapore came to play an important role in British naval strategy. The plan to construct an advanced facility in East Asia was first conceived in 1919. The Admiralty, with the support of Jellicoe, the First Sea Lord, as well as Lord Beatty, the Chief of Naval Staff, endorsed a plan to construct a base from which the RN could operate in the event of a conflict against Japan.[5] The proposal was accepted by the Cabinet in 1921, and thereafter, British East Asian strategy was based on the working assumption that upon the outbreak of hostilities, a fleet of capital ships would sail to Singapore.

Further impetus for developing a viable strategy to protect British interests in the region arose as a result of treaty obligations which Britain entered during the aftermath of the First World War. During 1921-22, the world's leading naval powers, including the US, Britain and Japan, held a conference at Washington, with the aim of securing peace and stability in East Asia by curbing the growing naval arms race and guaranteeing the territorial *status quo* in the Chinese mainland. Out of the host of agreements to emerge from the treaty, three of them had a direct impact on Britain's empire. The first of these was the naval limitation agreement, which capped the tonnage of the

---

[3] Nish, *The Anglo-Japanese Alliance*, 216-18.

[4] Nish, *Alliance in Decline*, 192-5.

[5] Ong Chit Chung, *Operation Matador: Britain's War Plans against the Japanese, 1918-41* (Singapore: Times Academic Press, 1997), 12-13; Kirby, *Singapore*, 6-7; W.David McIntyre, *The Rise and Fall of the Singapore Naval Base, 1919-1942* (London: Macmillan, 1979), 20-32.

USN, the RN and the IJN to a ratio of 5:5:3, respectively. As a result, Britain's centuries' old naval supremacy was emasculated, and without a large surplus fleet, the RN could no longer deploy significant vessels outside the home waters and the Atlantic Ocean. Second, the treaty forbade Britain from fortifying any of its bases north of Singapore, namely Hong Kong and Jesselton in North Borneo. The so-called 'non-fortification' clauses made it ever more imperative for the Royal Navy to possess an adequate base at Singapore. Third, the Anglo-Japanese Alliance was abrogated out of fear that its continuation might alienate the Americans, and replaced by the Four-Power Treaty signed between the US, Britain, Japan and China. The move caused much resentment from the Japanese, particularly within naval circles, who began to view the British Empire as a potential adversary.

Yet, whilst war against Japan became a distinct possibility, Britain did not face any pressing need to prepare for such scenarios. For this reason, a number of key questions concerning the RN's strategy were left up in the air. One of the questions concerned the time it would take for the fleet to sail to Singapore, and how long the 'period of relief' would last, during which time British ground forces would have to hold out on their own. Second, the strategy which the RN would follow once it arrived at Singapore was left open to further investigation. Last but not least, the date of completion for Singapore base was undecided. In the aftermath of the First World War, with virtually all of the Great Powers reverting to a policy of cutting back on defence spending and avoiding hostilities, Britain did not feel compelled to embark on any serious measures to bolster its military capabilities in East Asia.

British Retrenchment, 1922-31

For a large part of the decade following its inception, there were few efforts to press ahead with the construction of the Singapore naval base, mainly because Britain did not have to undertake measures to prepare for a large-scale war against Japan or any power for that matter, at least in the foreseeable future. At the international level, this was a period when world leaders sought to protect their national interests by relying upon the principle of collective security that was preached by the League of Nations, and to settle international disputes through diplomatic negotiation. British statesmen followed the credo that war was to be avoided at all costs. The Ten-Year-Rule stipulated that British strategy was premised on the assumption that there would be no major wars to fight for another ten years Last but not least, the economic losses of the last war led successive British governments to cut back on defence spending in order to provide adequate funds for reconstruction and public services. In East Asia, the British saw equally few reasons to substantially increase military commitments. Japan reverted to a phase of abstaining from territorial acquisition, and pursuing a policy of non-aggression. Tokyo's policies were dictated by a civilian-controlled government who advocated peaceful co-existence with the Western powers. Demands by navy and army circles for a more aggressive policy were quelled by the government, who insisted that Japan could not afford to risk an armed conflict with Britain or the US.

Under the circumstances, British leaders were most likely to take a lax attitude, and remain undecided over a number of key questions concerning the Singapore strategy. The only concrete decision made during the 1920s concerned the site where the base was to be constructed. In order to ensure ample space for storage and docking facilities, the base was to be built on the northern edge of Singapore island, away from the commercial heart and industrial sections of the colony. The size and composition of the coastal defences was left open to question. Indeed, the 1920s has been labelled as the 'wasted years'.[6] However, this was logical, in light of Britain's shortage of financial resources, coupled with the fact that it did not face any imminent threats to its position.

Strategic Overstretch, 1931-39

During the 1930s, the rise of Nazi Germany, coupled with Japan's pursuit of its expansionist policies on the Asian continent, confronted Britain with a multitude of threats against all parts of its empire, and defence leaders could no longer remain complacent about the possibility of a war breaking out in the foreseeable future. However, at the same time, the situation was somewhat complicated by fact that Britain lacked the resources to defend its worldwide empire against its potential enemies. After 1933, with Hitler's accession to power, followed by Germany's rapid rearmament, Britain needed to focus its attention on Europe. The economic effects of the Great Depression

---

[6] Kirby, *Singapore*, 11-18.

also placed substantial limits on defence spending and further jeopardized the prospects of Britain possessing any surplus forces for East Asia.

As a result, the 1930s saw the British downsizing their war plans. On one hand, there was significant progress towards the completion of the naval base at Singapore. In 1933, the Chiefs of Staff (COS) decided to complete the dockyards and the coastal defences, and the Cabinet agreed to release greater funds for the project.[7] The fixed defences, consisting of fifteen-inch and 9.2-inch calibre guns, were placed on the southern coasts of Singapore island, all provided with armour piercing shells that could be used against approaching ships. Munitions were also provided for use against land targets. The dockyard was completed in March 1938.

The completion of the base was not accompanied by any corresponding plans to dispatch a sizeable capital ship fleet in the event of war. The 1931 Naval War Memorandum, prepared by the Admiralty, stipulated that following its arrival at Singapore, the main fleet was to proceed to Hong Kong. British forces were to recapture the base in the event that it had fallen, and to prepare for attacks on positions closer to Japan's home islands which could provide bases for the blockading campaign.[8] In 1934, at a meeting of the COS, the representatives of the War Office and Air Ministry informed the Admiralty that they could not provide the necessary air and ground forces for operations against areas north of Singapore. As the decade progressed, British strategy turned to the defensive, until the feasibility of dispatching a fleet to Singapore itself became questionable. At the Imperial Conference of June 1937, the British delegation informed the Australian and New Zealand representatives that the size of the Far Eastern fleet would largely depend on the extent to which Britain was engaged in hostilities against Germany and Italy at the time.

The outbreak of the Sino-Japanese War in July 1937 raised further concerns about encroachment options of the regional *status quo*, and investigations of Britain's strategic situation produced an increasingly grim picture. In July 1938, the Admiralty's Director of Plans warned that it was not advisable for Britain to conduct a forward policy against

---

[7] Ibid., 28-32; McIntyre, *Singapore Naval Base*, 108-12.
[8] Christopher Bell, 'The Royal Navy, War Planning and Intelligence Assessments of Japan Between the Two World Wars', paper presented at *International Security Studies Symposium on Intelligence and International Relations*, Yale University, 3 May 1996, 8-9.

Japanese expansion, given the weak state of its capital ship fleet.[9] During the Tientsin Crisis of summer 1939, when the Cabinet was struggling to decide what action to take in response to the blockade of British concessions in China, Lord Chatfield, the Minister of Co-Ordination for Defence, lamented 'It is realized that the position in the Far East can only be strengthened with a corresponding weakening of [Britain's] position in Europe.'[10]

At the same time, the possibility of a Japanese invasion of Singapore was alleviated by visible indications which suggested that Japan's preoccupation with China, along with the economic and military costs of the campaign, had significantly limited its ability to undertake further territorial expansion. In early 1938, Dobbie, the GOC Malaya, and Percival, his chief of staff, warned that their examinations of Singapore's security had revealed that the defence of the entire Malay peninsula was vital, owing to the likelihood that the Japanese would establish bases on the peninsula for an overland advance towards the base. However, the War Office insisted that Japan was too preoccupied to embark on such a complicated expedition.[11] In September 1939, the COS went as far as to state that a prolonging of Sino-Japanese hostilities would in fact be advantageous, for the economic costs were most likely to prevent Japan from diverting its attention towards Britain's territories in Southeast Asia.[12]

Japan's attitude towards the western powers also showed a sincere willingness to avoid a confrontation that could place additional burdens on its resources. The government's prompt offer of compensation for the attacks on the US tanker *Panay* and HMS *Ladybird* in December 1937 gave rise to suggestions that the Imperial forces were most likely to respect British interests.[13] Intercepted Japanese diplomatic communications revealed that the foreign ministry was hesitant to agree to German calls for transforming the Anti-Comintern Pact into a formal military alliance, since such moves were likely to provoke Britain and the US into

---

[9] UKNA, ADM 116/4087 Minute by Director of Plans, Admiralty, 8 July 1938.

[10] UKNA, CAB 53/11 Annex I to Minutes of COS 300th meeting: Memorandum by Minister for Co-ordination of Defence to COS Sub-Committee, 16 June 1939.

[11] Louis Allen, *Singapore, 1941-1942* (Newark, DE: Delaware University Press, 1977), 44; Chung, *Matador*, 71.

[12] UKNA, CAB 80/3 COS (39) 52 Sino-Japanese Hostilities: Memorandum by COS, 28 September 1939.

[13] UKNA, WO 106/5364 Naval and Military Intelligence Summary No. 39, by General Staff, Shanghai, 9 July 1938; WO 106/5603 Extract from Hong Kong Intelligence Review No.5/38, 1 March 1938.

imposing sanctions.[14] Thus, despite the difficulties which Britain faced in defending its empire, the threat of a Japanese invasion was deemed minimal so long as the China incident remained unresolved.

British Strategy Restructured, 1939-41

During the period following the outbreak of the Second World War in Europe, Britain's strategic posture suffered a further deterioration. Preoccupied with the task of defeating Germany, London was unable to commit large forces to contain the Japanese, the latter starting to embark on moves which clearly undermined western interests in Asia. Nevertheless, in spite of the unfavourable circumstances, British defence officials maintained that their empire could be safeguarded, and that the Imperial forces were unlikely to attack Southeast Asia. Because the RN did not have the resources to dispatch a fleet to Singapore, the base was to be secured by nominally strengthening the ground defences of the adjacent Malay peninsula.

During the spring and summer of 1940, as Germany redrew the boundaries of Western Europe, Japan gained favourable opportunities to encroach on Britain's interests in Asia. The fall of France and Holland exposed Indochina and the Dutch East Indies. In June 1940, an agreement was secured with the Vichy government whereby the Japanese were granted access to bases in northern Indochina from which they could conduct long-range air raids against China's hinterland, where Chiang Kai Shek's nationalist government had relocated its headquarters. Japan also began to place pressure on Britain to cease supplying arms to China via the Burma Road. Most importantly, while Japan's ground forces were committed to operations in China, its navy was not extensively involved in the campaign, and thus able to deploy the bulk of its fleet for operations against western interests in the Asia-Pacific region.

To complicate matters, Britain's fixation with the war against Hitler restricted its ability to contain the Japanese. In August 1940, the COS warned that the cumulative effect of Germany's onslaught against the home islands and trans-Atlantic lifelines meant that even in an emergency, Britain could dispatch naval forces to protect Singapore

---

[14] India Office Library and Records, British Library, London (IOLR) L/WS/1/72 Weekly Circular by MI2 (War Office) for Military Intelligence Directorate (Army Headquarters India), 9 February 1939. MI2 was the section of the War Office's intelligence directorate that was responsible for Far Eastern affairs.

only with the greatest difficulty.[15] The newly appointed government of
Winston Churchill maintained that the British Empire needed to take
all possible measures to avert a war against Japan. Diplomatic negotiations were pursued with the other western powers who held imperial
possessions in East Asia, namely the Dutch government-in-exile and
the US, with a view to securing military cooperation against Japanese
aggression.[16] Yet, without adequate strengths, any effort to defend
Southeast Asia was unlikely to succeed.

Despite the mounting evidence which suggested that Britain faced
an ever-worsening position, defence planners did not believe that a
war against Japan would materialize. The complacency can be largely
explained by the fact that the available intelligence did not provide firm
indications that Japan intended to initiate hostilities with the western
powers, and under the circumstances, the British could only provide
vague assessments of how the situation in the Far East was likely to
unfold. For example, following the occupation of southern Indochina
in July 1941, the Far Eastern Combined Bureau (FECB), which was
the main body responsible for handling British intelligence activities in
Asia, went as far as to state, 'the primary objective... was to ensure that
Japan obtains all of the resources from Indochina', with the acquisition
of bases for southward expansion being secondary.[17]

The ambiguity surrounding Japan's strategy was compounded by
evidence which suggested that Tokyo was apprehensive about the possibility of intervention by the offstage superpowers, namely the USSR
and US. On the Asian mainland, as long a war against the Red Army
remained probable, the IJA needed to keep the bulk of its forces on
reserve in Manchuria. Nor did the nonaggression pact with Moscow in
April 1941 eliminate the prospect of hostilities. Germany's invasion of
Russia in June raised the possibility of the USSR collapsing, in which
case Japan's armies needed to be ready to advance into Siberia, so as to
share the spoils. On 16 July, the War Office disseminated the contents
of a decrypted diplomatic telegram which revealed that at a policy conference held earlier in the month, Japan's top leaders concluded that

---

[15] UKNA, CAB 80/15 COS (40) 592 Situation in the Far East in the Event of Japanese
Intervention: Report by COS, 15 August 1940.
[16] See Nicholas Tarling, *Britain, Southeast Asia and the Onset of the Pacific War*,
Cambridge: CUP, 1996, Chapters 2-3.
[17] UKNA, FO 371/27765 F 6949/9/61 Cipher No.21399: GSO 1 (Intelligence) to War
Office, 26 July 1941.

intervention in *Barbarossa* could not be ruled out.[18] Observations of increased troop concentrations on the Soviet-Manchukuo border also signified that contingency plans to intervene in the event of a German victory had not been abandoned.[19] As late as autumn 1941, the over-all consensus within Whitehall was that Japan's strategy was largely dictated by its need to prepare for hostilities against the USSR. The Joint Intelligence Committee (JIC), which acted as the chief producer of intelligence assessments for the British war cabinet, concluded that troop concentrations along the Manchurian border pointed to offen-sive action, and an attack on Singapore was not feasible unless the Soviets substantially reduced their strengths in East Asia.[20]

The Japanese also had to consider the possibility that an invasion of British and Dutch territories in Asia would entice the Americans into aiding their European allies. From late 1940, the Roosevelt admin-istration steadily departed from its isolationist policies and began to actively support Britain's war effort. The 'destroyers for bases' deal and Lend-Lease agreement committed the Americans to providing mate-rial help to their trans-Atlantic ally. In December, the US and Britain commenced negotiations for naval cooperation, and in March 1941, the ABC-1 agreement called for the two powers to coordinate their action in the event the US entered the war. When Japan occupied Indochina during 1940-41, Washington initiated the imposition of economic sanctions. In August 1941, at the Newfoundland summit, Churchill and Roosevelt signed the Atlantic Charter, which formed the basis for Allied war aims to rid the world of fascist aggression and secure national self-determination in the post-war world.

With the US playing a more vigorous role in world affairs, Japan was considered more likely to shun the military power which America could put up. The need to guard against US action was seen as a distinct obstacle to conducting large operations in the southern regions. In May 1940, the Director of Naval Intelligence concluded that, although Japan was capable of occupying the Dutch East Indies, the threat of US and British intervention negated any motives for such a move.[21] Ten days

---

[18] UKNA, WO 208/2259 War Office Weekly Intelligence Summary No.100, 16 July 1941.

[19] UKNA, FO 371/27838 F 7123/2822/61 Inter-Departmental Cipher No.0925Z: COIS Far East to Admiralty, 29 July 1941.

[20] UKNA, CAB 81/104 JIC (41) 362 Japan's Intentions: Report by JIC, 13 September 1941.

[21] UKNA, ADM 223/495 NID 0861 Japanese Intentions *vis-à-vis* the Dutch East Indies: Memorandum by J.H. Godfrey (DNI), 17 May 1940.

before Pearl Harbor, the JIC reiterated the conclusion it had reached back in January 1941, namely, that fear of a US attack on the home islands would prevent the IJN from dispatching sufficient capital ships to support an amphibious operation against the southern regions.[22] In the light of evidence which suggested that Japan faced difficulties in confronting the Allied powers, the British were likely to assume that the Imperial forces would move cautiously.

Aside from strategic complications, Japan's economic situation suggested that its ambitions were limited by the desire to avoid the consequences of sanctions. Tokyo's dependence on imported raw materials could only be taken as evidence that Western embargoes were bound to cripple its war making capacity. The possibility that sanctions might induce Japan to solve its problems by occupying new sources of raw materials in Southeast Asia was examined, but invariably dismissed. A report by the Axis Planning Section in April 1941 predicted that fears over economic reprisals were likely to limit Japan's actions to obtaining new supplies of resources through the peaceful annexation of areas which were beyond British and US control, including Indochina and Thailand.[23] In August 1941, after Japan occupied southern Indochina, the Joint Planning Subcommittee concluded that although the oil embargo and freezing orders imposed by the US and its western allies may have increased the possibility of war, fears over the effects of continued restrictions were more likely to 'make Japan pause and count the cost before taking another step forward.'[24]

On the eve of the outbreak of war, the Ministry of Economic Warfare (MEW) suggested that unless Japan made sufficient concessions to the western powers, it faced the prospect of suffering 'an economic situation as to render her ultimately unable to wage war, and reduce her to the status of second rate power.'[25] Although the Dutch East Indies were the only territories which could provide the required petroleum, the islands were not necessarily considered part of Japan's immediate

---

[22] UKNA, CAB 81/99 JIC (41) 11 Sea, Land and Air Forces Which Japan Might Make Available for Attack on Malaya: Report by JIC, 6 January 1941; CAB 81/105 JIC (41) 449 Possible Japanese Action: Report by JIC, 28 November 1941.

[23] UKNA, 81/101 JIC (41) 155 Future Strategy of Japan: Report by Axis Planning Section, 15 April 1941. The Axis Planning Section was a subsection of the JIC, whose task was to evaluate the strategic situation from the Axis powers' viewpoint and thereafter assess the moves they were most likely to pursue.

[24] UKNA, CAB 80/29 COS (41) 474 (Annex) Report by JPS, 3 August 1941.

[25] UKNA, WO 208/887 Japan – Economic Factors: Report by Ministry of Economic Warfare, 6 December 1941.

objectives. The British did not foresee how the presence of US and British forces in Asia, combined with the effects of the economic sanctions imposed in July 1941, would encourage, rather than dissuade the government and military elite to take a gamble.

For the purpose of strategic planning, the notion that the threat of Allied opposition could contain Japan led Britain to adopt a policy of relying on deterrence. Defence planners believed that they could compensate for their inadequate strengths in theatre by dissuading Japan from seeking further expansion. By 1941, the prevailing belief within Whitehall and Malaya Command was that if Britain undertook nominal improvements in its military position, Japan would lose confidence in the chances of conducting a successful invasion. Despite Britain's inability to dispatch a fleet to Singapore, the reorganization of Malaya's ground forces was viewed as a viable panacea.[26] One of the myths that has arisen in popular literature is that the British thought any attack on Singapore would come from the sea. Hence, they attempted to defend the base by relying upon the coastal batteries, and ignored the possibility that the Japanese could attack via Malaya. This myth arose mainly from stories circulated in the press following the fall of Singapore. However, during the decades after the end of the Pacific War, a number of historians have provided detailed accounts of how Britain did make a number of concerted measures to secure Singapore against a landward attack.

Among the most seminal works on the reorganization of Malaya's defences has been one written by O.C. Chung.[27] The main hindrance to the development of an adequate plan was not a lack of foresight on the part of British commanders, but the fact that Britain's strengths were tied down in North Africa and Europe. Between late 1940 and early 1941, British commanders, under the leadership of Sir Robert Brooke-Popham, the Commander-in-Chief (C-in-C) Far East, conducted a series of examinations of Singapore's security, and concluded that the possible establishment of Japanese bases on the Malay peninsula necessitated a revision of the base's defence system. In April 1941, following months of plea bargaining, the COS accepted Brooke-Popham's argument that Singapore's defence perimeter needed to be extended so as to include northern Malaya. The COS also approved preparations for a pre-emptive occupation of the Kra isthmus in the event of an invasion

---

[26] Chung, *Matador*, 142-69.
[27] Chung, *Matador*, passim.

(code-name Operation *Matador*). The plan involved a nominal increase in British strengths from twenty-six to thirty-two battalions. Churchill consented, on the condition that it did not entail any reduction of British strength in the home islands and the Middle East.

By late 1941, the prevailing belief among local commanders was that British forces in Malaya could contain the Japanese. The optimistic views held by the British were illustrated by a meeting held at GHQ Malaya during the autumn. The proceedings were described by B.H. Ashmore, a senior commander on Percival's staff, who recalled that the FECB representative 'painted a fairly indecisive picture', and was unable to provide much information whether the occupation of Indochina portended further moves.[28] Although the presence of enemy forces within proximity of Malaya raised worries, the upshot was that the Japanese high command would refrain from invading, after realizing that British beach defences and air opposition posed complications. Thus, the defence plan called for British forces to be concentrated at the most likely landing sites, and to hold successive lines of defence throughout the peninsula in an effort to forestall the invaders. Major-General Playfair, the chief of staff for GHQ Far East, went as far as to suggest that, even if the Japanese did invade, Singapore was unlikely to be endangered, so long as *Matador* was executed immediately and the enemy could be held at the beachheads.[29] In any case, the jungle could be relied upon to impede the Japanese advance.

A large part of the problem in formulating a realistic assessment of the threat posed by the IJA stemmed from the secrecy with which Japan had conducted its rearmament programme. This prevented the British from understanding how their adversary had become highly skilled at conducting amphibious landings and operations in jungle terrain, both of which contributed decisively to the Japanese successes. The Japanese army's efficiency was further discredited by its failure to bring its operations in the China theatre to a successful conclusion. Gordon Grimsdale, head of the FECB, recalled that the reverse at Taierchwang back in April 1938 formed the basis of Western beliefs that, since the Japanese were defeated by an inefficient Chinese army, the IJA itself

---

[28] Imperial War Museum, London (IWM), Percival Papers, P 49 Some Personal Observations of the Malaya Campaign, 1940-2: prepared by B.H. Ashmore, 27 July 1942.

[29] Liddell Hart Centre for Military Archives, King's College London (LHCMA), Brooke-Popham Papers 6/1/26 Most Secret Cable No.359/4: I.S.O. Playfair (Major-General of COS, GHQ Far East) to TROOPERS, 20 August 1941.

was in an even poorer state.[30] As post-mortems on the Malaya campaign admitted, the provision of accurate evaluations was hindered because the vast majority of Japanese troops in China were second-rate conscript soldiers who did not fully represent the IJA's capabilities.[31] Britain's underestimation therefore can largely be attributed to the lack of access to information on the efficiency of the armies which its forces were to encounter in Southeast Asia.

In the end, British ground forces proved incapable of withstanding the Japanese, and their weaknesses were clearly exposed when the invasion commenced in December 1941. The British did not possess any tanks, and their air force consisted mainly of obsolete aircraft. Troops were not trained to fight in rugged terrain and in conditions where they did not have access to reliable communications. Only one unit, the 1st battalion of the Argyll and Sutherland Highlanders, had any real training in jungle warfare. After landing in Malaya, it took the Japanese a little more than two months to capture Singapore. Numerous scholarly works, as well as first-hand accounts written by the commanders responsible for planning the defence of Singapore, have debated whether the capitulation was inevitable, and explored the question of how the available forces could have been deployed more effectively.[32] However, the final explanation must follow the post-mortem of General Sir Henry Pownall, commander of the British forces in the Far East, who concluded that commitments in Europe and the Middle East took priority, and preoccupations with matters closer to home negated the chances of protecting Britain's empire in Asia.[33] Because Britain was unable to divert its most effective forces to the Far East, Singapore had to be defended with insufficient strength.

---

[30] IWM CON SHELF (Grimsdale Papers) 'Thunder in the East': by G.E. Grimsdale, 1947, 10.

[31] UKNA, WO 208/1529 Extracts from a report by Lt.-Colonel Phillips, formerly GSO 1 (OPS) Malaya Command, 30 May 1942, and Report on Malaya and Singapore: drawn up by Major H.P. Thomas (OBE, Indian Army), 30 May 1942.

[32] Allen, Singapore, passim.; Henry G. Bennett, Why Singapore Fell (Sydney, AU: Angus & Robertson, 1944); Raymond Callahan, The Worst Disaster: the Fall of Singapore (London: Associated University Press, 1977); P. Elphick, Singapore: The Pregnable Fortress (London: Hodder & Stoughton, 1995); Kirby, Singapore, passim. and The War Against Japan, Vol. I, in series History of the Second World War (London: HMSO, 1957-70); A.E. Percival, The War in Malaya (London: Eyre & Spottiswoode, 1949); A.O. Robinson, 'The Malayan Campaign in the Light of the Principles of War' – Parts I and II, in RUSI Journal, Vol. 109, August and November 1964, 224-32 and 325-37; Ivan Simson, Singapore: Too Little Too Late (London: Leo Cooper, 1970); Alan Warren, Singapore, 1942: Britain's Greatest Defeat (London: Hambledon, 2002).

[33] LHCMA, Pownall Diaries, 25 February 1942.

Road to Disaster, August to December 1941

During the summer and autumn of 1941, as Britain faced an ever-increasing prospect of becoming entangled in a war against Japan, the war cabinet in London adhered to its policy of attempting to deter Tokyo from provoking hostilities. This objective was to be achieved through two courses of action, namely 1) a demonstration of Anglo-American solidarity, and 2) a nominal increase of British naval strength in the Far East. It was in these circumstances that Churchill decided to dispatch the capital ships *Prince of Wales* and *Repulse* on their ill-fated mission to Singapore.

By that time, as Japan's leaders were facing war options, Churchill's government maintained that a confrontation could be avoided. This was mainly because the British were counting upon America to discourage Japan, by making it clear that any act of aggression against the British Empire would elicit intervention. On one hand, Washington had emphatically stated that anti-imperialist opinion within the US government and public prevented America from deploying its forces to defend British territories, or even making any pledges to take military action in the event Japan moved into Southeast Asia. Nevertheless, policymakers in London maintained that, as long as Britain synchronized its policy with the US, and followed suit in imposing economic restrictions and hinting at further repercussions, the situation in the Far East could be stabilized. The occupation of southern Indochina did little to raise doubts that Japan would back down. During the last week of November, Churchill made repeated recommendations to Roosevelt for a joint Anglo-American declaration, on the belief that conquests could be prevented so long as Japan knew clearly that such moves entailed hostilities.[34]

By depending on the US to coerce Japan to behave more amenably towards the western powers, the British government surrendered the initiative of carrying out diplomatic activities, and made a minimal effort to negotiate directly with Tokyo. The British increasingly became mere observers of the international relations in East Asia and, consequently, they developed an erroneous view regarding the state of Japanese-American relations. In September, Tokyo ordered its

---

[34] Churchill College Cambridge Archives Centre (CCC), CHAR 20/44/117 and CHAR 20/46/2-3 Churchill's Personal Telegrams to Roosevelt, 5 and 30 November 1941, respectively.

ambassador at Washington, Nomura Kichisaburō, to commence negotiations with US Secretary of State Cordell Hull, in an attempt to secure a lifting of the economic sanctions. The Americans were adamant that the minimum price for rapprochement was a withdrawal from China. After five years of waging a costly war, the Japanese were bound to reject such terms. Yet, while the British were aware that the talks were not proceeding well, they were blissfully ignorant of how the US and Japanese positions were incompatible, to the point where a conciliation was impossible. As late as November, when the newly arrived hard line government of Tojo Hideki was finalizing Japan's war plans, the prevailing view was that confrontation could be avoided. GHQ India explained how, if Japan had to choose between commencing hostilities or sitting on the fence to await a favourable solution, past experience gave good reason to believe that Japan would take the peaceful option.[35]

In regard to military action, Churchill believed that a token British Fleet at Singapore would negate whatever optimism the Japanese had in their ability to wage war against an Allied combination. The view was reinforced by the fact that the US Pacific Fleet had been gearing up for mobilization at Pearl Harbor, and the leadership in Tokyo appeared hesitant about provoking a confrontation against an Anglo-American coalition. The idea of dispatching a deterrent naval force was conceived in August 1941. Admiral Sir Dudley Pound, the First Sea Lord, proposed that the battleships *Nelson*, *Rodney* and *Renown*, along with an aircraft carrier be dispatched to the Indian Ocean to 'deter the Japanese from sending any of their battleships to the area'.[36] All three battleships were vintage vessels, which did not possess the speed or armament to challenge the Japanese fleet. Churchill's response was that whilst the presence of obsolescent vessels was unlikely to help the RN's position in the event of hostilities, 'nothing would increase (Japan's) hesitation more than the appearance' of a British capital ship force.[37] Nobody in Churchill's Cabinet nor the Admiralty believed that a token force could resist a Japanese attack on Malaya. The intention was to show Japan that Britain still had a presence in that region. The British also had to take concrete measures that could reassure the Pacific Dominions,

[35] IOLR, L/WS/1/317 GHQ India Monthly Intelligence Summary No.11, Appendix C: Some Notes on Japan's position today, 3 November 1941.

[36] UKNA, ADM 205/10 Minute by Dudley Pound for Churchill, 28 August 1941.

[37] UKNA, ADM 205/10 Churchill's Personal Minute M.845/1 for Dudley Pound, 29 August 1941.

namely Australia and New Zealand, that London was at least making an effort to protect the empire.

During the following weeks, the idea of sending a token fleet to Singapore gained further support from influential members of Churchill's war cabinet, including Foreign Secretary Anthony Eden. At a Defence Committee meeting held in October, the decision was made to dispatch Force Z to Singapore, consisting of the modern battleship, *Prince of Wales* and vintage battle-cruiser *Repulse*.[38] The British leadership's adherence to its illusionary faith, that Japan could be deterred, was further underlined by Pound's letter to the British Admiralty Delegation at Washington, in which he suggested that the arrival of the capital ships at Singapore may well have made the Japanese hesitate.[39] Given the fact that war was bound to result in an ultimate defeat for Japan, combined with evidence which suggested that its actions were being limited by its apprehension about facing a costly confrontation with the Allies, Britain was most likely to adhere to its belief that it could avert hostilities. Under the circumstances, it was logical to give minimal consideration to any substantial enhancement of Britain's defensive capabilities *vis-à-vis* Japan. Only the speed and scale of Japan's victories in Southeast Asia could convince the British that their adversary could not be dissuaded.

In the end, Force Z was sunk by enemy bombers, off the coast of Malaya, two days after the outbreak of the Pacific War, while attempting to intercept Japanese transports heading for the coast. The loss of the ships can be attributed to a number of errors made by Admiral Philips, the commander of Force Z. Among the most notable mistakes was to dispatch the vessels when reliable fighter cover could not be guaranteed. However, even this decision has to be viewed in light of a lack of any concrete information concerning the capabilities of Japan's naval air services. Japanese security arrangements prevented the British from gaining information on how their adversary had made

---

[38] UKNA, CAB 69/2 Cabinet Defence Committee (Operations) 65th meeting, 17 October 1941. The most comprehensive secondary accounts on this aspect include: Ian Cowman, 'Main Fleet to Singapore? Churchill, the Admiralty and Force Z', in *Journal of Strategic Studies*, Vol. 17, 1994:2, 79-93; John Pritchard, 'Churchill, the Military and Imperial Defence in East Asia', in Saki Dockrill (ed.), *From Pearl Harbor to Hiroshima: the Second World War in Asia and the Pacific, 1941-45* (Basingstoke: Macmillan, 1994), 26-54; also, Chapter 11 of Stephen W. Roskill, *Churchill and the Admirals* (London: Collins, 1977).

[39] UKNA, ADM 205/9 Personal Letter from Pound to Little (British Admiralty Delegation, Washington), 6 December 1941.

meticulous efforts to develop the tactics needed to destroy enemy forces at sea. In a similar manner to how British officials disparaged the IJA, their opinions of the IJN showed contempt, and focused on how the Japanese had yet to combat opponents with advanced technological resources and tactical skill. For example, the Admiralty's Naval War Memorandum for the Far East, prepared in winter 1939, surmised that, while total British and Dominion air strengths were lower than their Japanese counterpart, there was little reason to fear the latter's combat capabilities.[40] Owing to their inexperience dealing with anti-aircraft fire, and lack of training in the functions of torpedo bombing, oversea reconnaissance and target spotting, the Imperial forces were judged to be inferior in these fields.

In addition to technical shortcomings, the IJN air arm was considered likely to face logistical problems in supporting long range operations across the expanses of ocean in Southeast Asia. This was a potential limit to its ability to conduct operations against Malaya. As early as 1937, the COS stated that even if a seventy-day window was available before the arrival of the main fleet at Singapore, a Japanese invasion was still unlikely, due to the difficulties involved in bringing sufficient air power over long distances.[41] The establishment of bases in Malaya was deemed equally unlikely, given the absence of suitable anchorage sites on the east coast, along with the fact that the development of bases on the west coast required the protection of long lines of communication. In assessing the forces which the Japanese could bring to bear, prepared in January 1941, the JIC reassured that even in the event of an occupation of bases in southern Indochina, Singapore was only within range of heavy bombers, while light bomber attacks required the establishment of bases in the Kra isthmus.[42]

Similarly, the Air Ministry predicted that in addition to shortages of aircraft with adequate ranges, maintenance difficulties and wastage resulting from operations over large stretches of water were likely to limit the scale of attack.[43] Correspondence between Brooke-Popham and the COS suggested that fears of the possible scale of air attack were

---

[40] UKNA, ADM 116/4393 Naval War Memorandum (Eastern), Undated (?? February 1939).

[41] UKNA, CAB 53/31 COS 596 Appreciation of the Situation in the Far East, 1937: Report by COS, 14 June 1937.

[42] UKNA, CAB 81/99 JIC (41) 11 Sea, Land and Air Forces Which Japan Might Make Available for Attack on Malaya: Report by JIC, 6 January 1941.

[43] UKNA, AIR 22/75 Air Ministry Weekly Intelligence Summary No.103, 20 August 1941.

confined to local commanders, whose warnings were dismissed as unduly alarmist.[44] The decision to dispatch the *Repulse* and *Prince of Wales* on their mission was admittedly based on scanty and misleading intelligence on the maneuvres which the Japanese air services could carry out.[45] In his recollections of the Far Eastern conflict, Hillgarth, who served as the Chief of Naval Intelligence in the Southeast Asia theatres, admitted that the gravest miscalculation regarding the force attacking Malaya was not on its strength and location, but the quality of the bomber convoy.[46]

Conclusions

The fall of Singapore was, and still remains, as Winston Churchill described it, 'the worst disaster and largest capitulation in British history'. [47] Nor is there any debate as to whether British strategy against Japan during the years leading to the Pacific War was inadequate; the Japanese invasion of Malaya caught the British completely unprepared. However, the decisions made by British politicians and defence planners have to be viewed in light of the fact that Britain simply did not have the financial resources nor military strength to defend its global empire against all of its enemies. Throughout the interwar years, British grand strategy was shaped primarily by the need to contain Germany's threat to the home islands. The second major concern was to protect Britain's lifelines in the Atlantic and Mediterranean, as well as its vital Middle Eastern oil supplies. The defence of Malaya and Singapore was lowest on the list of priorities. As the 1930s unfolded, the overall build-up of armed forces in Europe and Japan's own effort to enhance its capabilities reduced the margins of British military supremacy.

---

[44] LHCMA, Brooke-Popham Papers 6/1/6 Telegram from COS to Brooke-Popham: Signal X776, 10 January 1941; and 6/1/11 GHQFE/494: Brooke-Popham to Air Ministry, 1 March 1941.

[45] CCC, DUPO 5/5 Loss of the HMS *Prince of Wales* and *Repulse* on 10 December 1941: by Training and Staff Duties Division (Historical Section), August 1948; Richard Hough, *The Hunting of Force Z: The Sinking of the Prince of Wales and Repulse* (London: Collins, 1963), 117.

[46] UKNA, ADM 223/494 *Pearl Harbor and the Loss of the Prince of Wales and Repulse*: by Captain Hillgarth, Royal Navy, 1946.

[47] Winston S. Churchill, *The Second World War, Volume IV: The Hinge of Fate* (London: Cassell, 1951), 81.

Britain's strategy in the Far East prior to 1941 was determined mainly by its shortage of surplus resources, and preoccupations with matters closer to home diminished the prospects of securing its empire against Japanese aggression. The naval agreements of the 1920s had reduced the strength of the RN, Britain's primary tool to defend the Empire and in the 1930s, rearmament required the service to make hard choices. In this respect, the available intelligence did not suggest that Japan had either the capacity or willingness to provoke a war against Britain and its western allies; on the contrary, it provided convincing reasons to conclude otherwise. Under the circumstances, it was logical for the British to develop a complacent attitude about the possibility of Japan launching a successful invasion of Singapore.

PART TWO

STRATEGIC PRIORITIES FROM THE COLD WAR TO IRAQ

BALANCING THREAT PERCEPTIONS AND STRATEGIC PRIORITIES: JAPAN'S POST-WAR DEFENCE POLICY

Noboru Yamaguchi

Japan's rapid recovery from the devastation of the Second World War has been regarded by many analysts as a miracle. One of the key factors that favoured such an impressive recovery was Japan's ability to focus on the reconstitution of economic rather than military power. In the early 1950s, the country's political leadership devised a national strategy based on what later became known as the 'Yoshida Doctrine', from the name of Prime Minister Yoshida Shigeru – its primary architect. This strategy combined the prioritization of economic measures to favour recovery and growth with limited rearmament, which reflected the requirement of the new Constitution, and a security policy relying on an alliance with the United States.[1]

Defence policy did not rank on a par with economic measures in the government's agenda, but Japanese strategic planners consistently sought to develop sound policies with the limited means they were provided to meet the nation's core priorities. This chapter examines the evolution of Japan's defence policy in relation to changes in threat perceptions throughout the Cold War, and into the initial years of the post-Cold War period. The chapter argues that in the Cold War, Japan's threat perceptions evolved in nature, prompting a transformation of the purpose of Japan's military apparatus but not the acceleration of the pace of rearmament. Japan's defence build-up began in 1950 when the Korean War broke out. Threat perceptions focused on domestic instability, an instability caused by devastated economic and social

---

[1] For a brief analysis of the Yoshida Doctrine, cf. Christopher W. Hughes, *Japan's Re-emergence as 'Normal' Military Power* (Adelphi Paper N.368-9, Oxford: Oxford University Press, 2004), 21-40. Yoshida's views whilst they were never intended to become a 'formal' doctrine heavily informed the choices of subsequent Prime Ministers. Nakajima Shingo, *Sengo Nihon no Bōei Seisaku. 'Yoshida Rosen' wo Meguru Seiji, Gaikō, Gunji* (Japan's Post-War Defence Policy. Politics, Foreign Policy and Military Affairs about the 'Yoshida's Line', Tokyo: Keiō Gijuku Daigaku Shuppankai, 2006), 5-14.

infrastructures along with fear of the spread of communism. As the Cold War intensified and the Soviet Union built up its armed forces in the region, Japan's defence posture progressively shifted its focus from domestic to external threats.

The chapter further examines how, by the end of the 1980s, the evolution of Japan's strategic priorities required the country to reconsider its defence posture. Since the reconstitution of its military forces, Japan had refrained from dispatching men and assets overseas, with one exception during the occupation period, when Japanese minesweepers were sent to support United Nations forces fighting on the Korean peninsula. After the Cold War ended, the involvement in world affairs and international security entered the grammar of Japan's defence planners. Since the early 1990s, the balance between homeland defence, based on the regional threat perception, and strategic requirements to participate in international peace operations has been a central issue in the debate over Japan's defence and security policy.

Early Cold War Threat Perceptions and the Defence of the Archipelago

At the end of the Second World War in August 1945, Japan was completely de-militarized by the occupying Allied forces led by General Douglas MacArthur. The major security concern in this region in the late 1940s was the security against Japan, the defeated former enemy. Under the Allied occupation and with guidance from MacArthur's headquarters, the Constitution of Japan was established in 1947. The Allied powers' primary goal for the Constitution was to restrain Japan from becoming a regional threat again. The Constitution uniquely renounced war and imposed self restraint from rearmament, as expressed in Article 9 that reads as follows:

> Aspiring sincerely to an international peace based on justice and order, the Japanese people forever renounce war as a sovereign right of the nation and the threat or use of force as means of settling international disputes.
>
> In order to accomplish the aim of the preceding paragraph, land, sea, and air forces, as well as other war potential, will never be maintained. The right of belligerency of the state will not be recognized.

'The Switzerland of the Pacific', as noted by General MacArthur in 1949 was the key expression for capturing Japan's neutrality in any military conflict.[2] The turning point came as the Korean War broke out on 25 June 1950. As the four US Army divisions stationed in Japan were deployed to Korea, the 'defence of Japan' as opposed to 'security against Japan' became the issue. On 8 July, the Allied headquarters ordered the Government of Japan to create the National Police Reserve (NPR) to assume domestic security missions formerly conducted by US forces. By the end of the year, a 75,000-strong NPR was activated. It was later reorganized into the National Safety Force in 1952 which in turn acted as the forerunner of the Japan Self-Defence Forces (JSDF).

At the beginning of Japan's rearmament, internal threats caused by domestic instability were a paramount political concern. Despite the breakout of the Korean War, Japan's leaders including Prime Minister Yoshida seemed to have more concerns over domestic security than the defence of Japan against a potential spill over from the war on the Korean peninsula. The creation of the NPR was repeatedly explained as a measure to deal with threats against stability at home. In the late 1940s, the rise of communism in East Asia seemed prominent, and in Japan social and economic disturbance, public insecurity caused by terrorism, and political infiltration through ideological propaganda were perceived as major sources of threat. Measures to deal with such internal threats included the strengthening of domestic security, the prosperity of democratic institutions, the increase of economic initiatives, and counter-propaganda. This was no easy task. The country was vulnerable to communist forces. It had a weak position in the international community. It had to reintegrate the approximately six million Japanese nationals repatriated from abroad into the fragile social and economic infrastructures of the country. Further, its leadership was still very divided over the debate on who should be held responsible for the drama of the war.[3]

The 'Yoshida Doctrine' was the appropriate answer to Japan's problems of stability and indeed, it led the country to a successful reconstruction and economic recovery. Prime Minister Yoshida had a strong belief that Japan should limit military spending due to the devastation

---

[2] Akihiko Tanaka, *Anzen Hosho – Sengo 50 nen no Mosaku* (Security: Examining the 50 years After the War, Tokyo: The Yomiuri Shinbun-sha, 1997), 38.

[3] Hiroshi Nakanishi, *Haisenkoku no Gaiko Senryaku – Yoshida Shigeru no Gaiko to Sono Keishosha* [*Diplomatic Strategy of a Defeated Nation: Shigeru Yoshida and His Successors*], presented at the second NIDS International Forum on War History on 15-16 October 2003.

caused by the war in the Pacific in order to rebuild its economic base and rely upon the alliance with the United States in terms of the defence of Japan. In addition, Yoshida was convinced that one of the major reasons for Japan to venture into a war that destroyed the archipelago was Japan's economic weakness in the 1930s. Therefore, the economy was his highest priority and rearmament was to be conducted to prevent it from affecting development. Recent studies on Yoshida's policy point out that his 'doctrine' was not born out of naïve idealism, but was grounded in the realm of realism. He believed that if Japan possessed military forces with expeditionary capabilities, this might have led to domestic political disputes and international concerns over Japan's military capabilities among nations in Asia, Oceania and Europe.[4] In realist terms, the policy of a lightly armed Japan worked as a positive element by trying to avoid a security dilemma within the international community.

At the outset, Japan's rearmament was meant to secure its own territory against internal threats to social and political stability, with the NPR being primarily a constabulary organization. External operations were out of its scope.[5] Yet, throughout the 1950s, this domestic focus evolved for two reasons. The Berlin Blockade in 1948-49 and the Korean War in 1950-53 signalled an intensification in the tones of the Cold War, and the 'communist threat' became apparent in both the European and the Pacific theatres. Concurrently, Japan's economy received a significant stimulus, thanks in particular to the need for supply and maintenance by the United Nations Forces during the Korean War. As a result, Japan's security concerns over domestic instability caused by devastated economic and social infrastructures were gradually eased. Japan's perception of external threats became more serious than internal threats.

When the Japan Defence Agency (JDA) was established and the JSDF were established in 1954, a serious debate occurred within political circles over the question of whether the highest priority for the new

---

[4] Ibid.

[5] By order from the headquarters of the Allied forces, Japan sent forty-six mine sweepers to the Korean waters and demolished twenty-seven mines which resulted in one missing in action and eighteen wounded from October to September 1950. Since Japan was under Allied occupation and had no military organization, MCM operations were conducted by a civilian organization, the Maritime Safety Agency which belonged to the Ministry of Transportation and sailors wore uniforms and hats of civilian officials. James E. Auer, *The Postwar Rearmament of Japanese Maritime Forces, 1945-71* (New York, London: Praeger, 1973), 64-8.

institutions was to be the defence against external aggressions.[6] Eventually, the debate favoured this argument and defence against direct aggression was listed at the top of the missions of the new military organization. As stated in the JSDF Law of 1954, 'in order to protect the peace and independence of Japan and preserve its safety, the (J)SDF are charged with the main duty of defending the nation against both direct and indirect aggression (large-scale internal revolt and rioting caused by intervention or incitement by a foreign country)'.[7] The JSDF was expanded during four multi-year plans, known as Defence Build-up Programmes, stretching over a period from 1958 to 1976. The main objective of those plans was to build up a basic critical mass of 'efficient defence capabilities to most effectively deal with aggression from the level of local military conflict by conventional weapons'. The emphasis on a progressive growth focusing on quantitative aspects can be easily detected in the trends summarized in Table 4.1.

As designs for Japan's post-war military power were taking shape, the question of the 'size' of Japan's defence forces was a serious subject of discussion between the Government of Japan and the American Government as well. The two governments held different views on the subject. In 1953, Assistant Secretary of State Walter Robertson proposed personnel strength of 325,000 organized into ten army divisions, whereas Financial Minister Ikeda Hayato, the government's special envoy, insisted that a similar force structure could be achieved with a

Table 4.1. Japan's Defence Build Up (1952-6)

	NSF/GSDF	MSDF	ASDF
	Active Strength	Total Tonnage	Combat aircraft
FY 1952	110,000		
FY 1960	170,000	112,000 tons	1,133
FY 1966	171,500	140,200 tons	1,095
FY 1971	179,000	144,000 tons	940
FY 1976	180,000	198,000 tons	840

Source: Asagumo Shinbun-sha, *Bōei Handobukku* (Handbook for Defence – Tokyo: Asagumo Shinbun-sha, 2008), 74-9.

---

[6] Tanaka, *Anzen Hosho*, 132.
[7] Japan Defence Agency, *Nihon no Bōei* (Defense of Japan – Tokyo, translated into English by The Japan Times, Ltd., 1987), 79.

target goal of 180,000 troops. Ikeda's argument was based on the notion that as Japan's defence force was to operate within its own territory, it would be supported by domestic infrastructures requiring minimum additional capabilities for logistical support. Such capabilities precluded the option of using Japan's ground forces in expeditionary mission for overseas deployments. Provided Japan's constitutional, economic and political agendas of the time, a consensus was reached and the force level of the JGSDF was set to a personal strength of 180,000 with fairly limited combat service support units.

The Basic Defence Force Concept: From Détente to the End of the Cold War

By the mid-1970s, tensions between the Western and Eastern blocs had eased in a mood of *détente* that heavily impacted Japan's threat perception. The Soviet Union and the United States had a series of Strategic Arms Limitation Talks, or SALT, the first in 1972 (SALT I) and a second in 1979 (SALT II). In Europe, several agreements were reached between West and East, including the Helsinki Accords signed at the Conference on Security Cooperation in Europe held in 1975. *Détente* seemed to be extended to the Asia-Pacific region where President Richard Nixon visited the People's Republic of China and normalized relations with the country in 1971. Japan followed the United States by normalizing its relationship with China in 1972, whilst the United States began its withdrawal from Vietnam after the Paris Peace Accords in 1973.

Throughout this period, in Japan, the four multi-year plans had offered steady guidance for the strengthening of the JSDF, and the force levels initially sought in the 1950s was almost achieved by the second half of the 1970s. In 1976, a new concept, called 'Basic Defence Force' was adopted by the government and articulated into the National Defence Programme Outline (NDPO 1976) which came to represent the primary reference in matters of defence policy for the rest of the Cold War. This concept was based on an assumption that 'the international political structure in this region – along with continuing efforts for global stabilization – will not undergo any major changes for some time to come', and that there seemed 'little possibility of a full-scale military conflict between East and West, due to the military balance – including mutual nuclear deterrence – and the various efforts being made to stabilize international relations'.[8] The NDPO 1976 focused on the ability to cope effectively with situations up to the point of limited and small scale aggression and set national strategy as follows:

> Should direct aggression occur, Japan will repel such aggression at the earliest possible stage by taking immediate responsive action and trying to limit a small-scale aggression, in principle, without external assistance. In cases where the unassisted repelling of aggression is not feasible, due to scale, type or other factors of such aggression, Japan will continue an unyielding resistance by mobilizing all available forces until such time as cooperation from the United States is introduced, thus rebuffing such aggression.[9]

In short, the 'Basic Defence Force' concept sought to maximize Japan's own efforts for territorial defence in a climate of *détente*. In that context, whilst it was difficult to predict the evolution of super powers' relations and their impact on Japanese defence policy, this very incertitude posed questions on the identification of an appropriate amount of military expenditures to meet the Soviet challenge in the Pacific, Japan's main conventional threat in the region. Thus, the concept was regarded among officials in defence circles as 'non-threat based'. In addition, the Oil Shock in 1973 put Japan's economy under considerable strain, with consequent attempts by political authorities to limit defence expenditures. Those circumstances allowed Prime Minister Miki Takeo to set the limit on defence spending to 1% of Japan's GDP in 1976. This ceiling remains intact to date, with the only exception in 1987 when Prime Minister Yasuhiro Nakasone managed to breach it.

Nakasone's initiative took place in a security environment very different from the détente. Indeed in the 1980s, the non-threat-based approach soon lost its validity as tensions intensified again. After the Soviet invasion of Afghanistan in support of the Marxist government in December 1979, it became apparent that the United States would not ratify the SALT II treaty. Washington and a number of its allies as well as China (in an increasing confrontation against the Soviet Union) decided to boycott the Summer Olympic Games held in Moscow in 1980. In 1979, NATO made the so-called 'Double Track Decision', deploying the *Pershing II* Medium-Range Ballistic Missiles (MRBMs) and Ground-Launched Cruise Missiles (GLCMs) in Europe

---

[8] Japan Defence Agency, *National Defence Programme Outline* (Tokyo, 1976). The NDPO 1976 was adopted on 29 October 1976 by the National Defence Council, and approved on the same day by the Cabinet.
[9] Ibid.

in response to the Soviet deployment of SS-20 MRBMs in the area west of the Ural Mountains.

In East Asia, the Soviet military build-up in the region drew the attention of Japan's defence planners, a consideration that was reflected in the focus that this menace occupied in Japan's Defence White Papers throughout the 1980s. As late as 1987, the *Defence of Japan* noted that 'the (Soviet) ground forces have increased steadily since 1965. Of the approximately two million ground troops of the entire Soviet land-based forces which are grouped into 211 divisions, approximately 500,000 troops in 57 divisions are currently stationed near the Sino-Soviet border. In the Far East (roughly east of Lake Baikal) about 390,000 troops, or 43 divisions, are deployed at present.'[10] Other developments that attracted Tokyo's concerns included Soviet strategic, naval and air forces. The air and missile components of Soviet military capabilities rapidly increased in intermediate–range nuclear forces, including some 85 *Backfire* medium-range bombers and some 170 SS-20 MRBMs deployed in central Siberia and the area around Lake Baikal. The Soviet Pacific East Fleet underwent a major built-up, with the introduction of Delta III ballistic missile submarines (SSBN), and sophisticated combat surface ships such as the *Kiev* class of aircraft carriers, the *Kirov* class of nuclear-powered guided missile cruisers as well as the *Sovremenny* and *Udaloy* class of guided missile destroyers. In the air, the Soviet Union could count on its latest fighters such as Mig-23/27s, Mig-31s and Su-27.

As a result of this massive increase in firepower, throughout the 1980s, domestic debates over Japan's defence posture focused on what capabilities were necessary for the country to qualitatively enhance its effectiveness. The strengthening of the Japan-US alliance became a core issue in Japan's security debate and a primary means for Japan to increase the effectiveness of its limited military power. The United States had since the early 1980s repeatedly asked Japan to increase its defence spending and to assume a greater share of responsibility for the defence of the western Pacific. In particular, the US government considered that this burden sharing could concentrate in Japan's case on the ability to protect air and maritime spaces around Japanese waters and up to 1,000 nautical miles along the main Sea Lanes of Communication (SLOCs). Within this set of missions, in a policy planners' meeting between representatives of the two governments held in Hawaii in June 1981,

---

[10] Japan Defence Agency, *Nihon no Bōei 1987*, 38-48.

particular emphasis was given to Japan's capability to cope with Soviet submarines and Backfire bombers.[11]

Prime Minister Yasuhiro Nakasone, who led the cabinet for five years from 1982, played an important role in both promoting the alliance and enhancing the JSDF's capabilities. In 1983, in his response to American requests, Nakasone declared that Japan would become an 'unsinkable aircraft carrier' to prevent Soviet Backfire bombers from freely operating beyond the Japanese territories and would secure Japan's three main straits of strategic importance (Soya, Tsugaru and Tsushima).[12] Since Soviet fighters, bombers, surface combat ships and submarines stationed around Vladivostok had limited ways of deploying to the Pacific either by flying over the Japanese territory or navigating through one of the three straits, the defence of the Japanese archipelago in itself had crucial geo-strategic significance.[13] This point was eloquently stated by a Prime Minister's advisory group summoned in an effort to develop a post-Cold War defence policy in 1994. As the report of the advisory group put it, 'the defence capability of Japan in the Cold War period was built up and maintained for the primary purpose of preparing for the attacks on Japanese territory by hostile forces', and 'Japan's mission was to defend the country based strictly on the right of self defence. In light of its geographical position, however, Japan naturally played an important role in the ant-Soviet strategy of the Western bloc'.[14]

In a context of military operations, whilst the defence of maritime communications took a leading role in national defence, ground operations still remained a major concern for Japanese planners. The amphibious capabilities of the Soviet Far East Forces remained remarkable, and Japanese fears were shared by European observers too, with *Jane's Fighting Ships 1983-1984* voicing that 'The most likely course of

---

[11] Tanaka, *Anzen Hosho*, 289-91.

[12] Hidetoshi Sotooka, Masaru Honda, Toshiaki Miura, *Nichibei-doumei Hanseiki – Anpo to Mitsuyaku* (A Half Century of Japan-U.S. Alliance: Security and Secret Arrangements – Tokyo: Asahi Shinbun-sha, 2001), 379-81.

[13] On Japan's efforts in sealing the archipelago's three straits, see Alessio Patalano, 'Shielding the "Hot Gates": Submarine Warfare and Japanese Naval Strategy in the Cold War and Beyond (1976-2006)', *The Journal of Strategic Studies*, Vol. 31, 2008:6, 859-95.

[14] Advisory Group on Defence Issue, *Nihon no Anzen Hosho to Boeiryoku no Arikata – Nijuu Isseiki he Mukete no Tenbou* (The Modality of the Security and Defence Capability of Japan – The Outlook for the 21st Century – Tokyo, August 1994, hereafter 'Higuchi Report'), 11. The Advisory Group was first summoned by Prime Minister Hosokawa on 28 February 1994 and submitted the report to Prime Minister Murayama on 12 August 1994.

action in a time of war would be for the considerable Soviet amphibious forces to occupy northern Hokkaido, thus ensuring reasonable cover for their naval deployment.'[15] By the mid-1980s, the Soviet Union had deployed its only Naval Infantry Division in the Far East. This was supported by an amphibious fleet equipped with *Ivan Rogov* Class amphibious assault ships and the *Ropucha* class landing ships as well as other vessels. Japanese strategic thinkers recognized that the Soviets would employ such forces to pursue a possible strategy for protecting its SSBNs deployed in the Sea of Okhotsk. As the range of Soviet submarine launched ballistic missiles (SLBMs) was extended to be capable of covering the continental United States from areas close to the Soviet coast, the Sea of Okhotsk and the Barents Sea became possible bastions for Soviet SSBN forces.[16] For such a strategy, the seizing of northern Japan to assure the bastion in the Sea of Okhotsk became a more realistic scenario than before. This notion reinforced Japan's concerns over Soviet military capabilities and added to the necessity to accelerate JSDF modernization programmes. By the end of the 1980s, the introduction of new combat platforms and systems such as P3-C maritime patrol aircraft, the F-15 fighter aircraft, the SSM-1 land-based anti-ship missiles along with surface combat ships and air defence were all part of a wider attempt to neutralize the Soviet naval and amphibious threat.

Post-Cold War Threat Perception Adjustments and Changes in Strategic Priorities

Whilst Japan's defence policy underwent major changes in the years following the Gulf War, the core impetus for this was the change in the strategic environment brought about by the end of the Cold War. During the Cold War, in Europe, NATO countries could count on the multilateral alliance to form a unified front against the massive land forces of the Eastern powers. In East Asia, Japan stood face to face with the far-eastern military district of the Soviet army separated only by a maritime border. For the Soviet Pacific Fleet, the Soya, Tsugaru and Tsushima Straits were the gateways to reach the East China Sea or the Pacific Ocean via the Sea of Japan from its base in Vladivostok. Hence,

---

[15] John Moore (ed.), *Jane's Fighting Ships 1983-1984,* (London: Jane's Publishing Company, 1983), 139.

[16] Shigeki Nishimura, 'Nihon no Bōei Senryaku wo Kangaeru' (Thinking about Japan's Defence Strategy), *Shin Bōei Ronshū* (*New Defence Annals*), 50-79.

the protection of the Japanese mainland, which controlled these important straits, was an important contribution to the overall objectives of the West's containment strategy of the Soviet Union.

All this changed with the demise of the super powers confrontation. Almost overnight, Japan's contribution to the West through its Cold War defence posture became decisively less important. The geo-strategic importance of Japan dropped significantly due to the elimination of an anti-Soviet strategy. In that respect, in the early years of the Post-Cold War era it became necessary for Japan, as a nation which boasted one of the largest economies in the world and enjoyed the benefits of international peace and prosperity, to search for a new path to contribute to the international community. University of Tokyo Professor Akihiko Tanaka remarked on this subject that, 'whilst it is paradoxical, there was no need for Japan to become aggressively involved in the world's international conflicts during the Cold War. But the end of the Cold War served to place the world at a distance, and this became an issue for Japan's security.'[17]

The Gulf War in 1991 was an eye opening experience for Japan's security and defence policy planners. The Japanese government provided approximately 13 billion dollars. This represented almost 20% of the total financial burden of the United States and its allies making Japan the third largest financial contributor after Saudi Arabia and Kuwait.[18] However, this contribution by Japan was untimely and given out incrementally, and Japan was judged as doing too little too late. More crucially, the financial effort was not integrated with a contribution in personnel, a fact that was not received with approval by the international society.[19] In March 1991, the Kuwaiti government purchased a one-page space on major American newspapers like *The Washington Post* and *The New York Times* to express gratitude to nations which had made an effort towards the liberation of Kuwait; Japan was not included in the list of contributing nations. As one American friend commented to this author, 'Someone who simply writes checks in the office while

---

[17] Tanaka, *Anzen Hosho*, 309.

[18] According to reports by the US Department of Defense, the international community contributed approximately 88% (54 billion dollars) of the total outlay (61 billion dollars) during the Gulf Crisis and Gulf War.

[19] In November 1990, the Diet failed to adopt the 'Act on Cooperation for United Nations Peacekeeping Operations and Other Operations' submitted by Prime Minister Kaifu Toshiki. This proposed law aimed to allow participation in United Nations activities for peace such as United Nations peacekeeping operations or UN-authorized coalition activities.

the rest of us are working hard in the store cannot be called a partner.' Japan's 'cheque book diplomacy' proved to be highly ineffective.

This experience left a deep mark on Japanese policy planners. One of the primary responses to these events was the consideration to use Japan's defence forces for international peacekeeping missions. Immediately following the end of the Gulf War, the Japanese government made the decision to deploy minesweepers in the Persian Gulf in postwar de-mining activities. Subsequently, after an intense debate in the Diet, in 1992 Japan adopted the Act of Cooperation for United Nations Peacekeeping Operations and Other Operations which authorized the participation by the JSDF in United Nations peacekeeping operations. Shortly after the new law was enacted, the JSDF were dispatched to take part in UN-led peacekeeping operations and international disaster relief missions in Cambodia (1992-93), in Mozambique (1992-95), in the Golan Heights (1996-present), in Honduras (1998), and more recently, in East Timor (2002-2004).

The international peace cooperation activities began against this background, but the actual JSDF deployments were carried out with considerable restraint and prudence. For instance, the primary tasks of the international peacekeeping force (oversight of disarmament, encampment and patrol activities) stipulated by the 1992 Act of Cooperation for United Nations Peacekeeping Operations and Other Operations were put on hold. This constrained the actions that Japanese troops could conduct on the ground and the final restrictions were lifted in December 2001 as a result of the domestic and international understanding gained from the performance and experience during six overseas deployments over the previous nine years.

With regard to this prudent stance, the *Defence of Japan 1992* well summarized the discussions surrounding the establishment of the Cooperation for United Nations Peacekeeping Operations and Other Operations Act. First, it was necessary to guarantee that the JSDF participation in the UN peacekeeping operations 'were not (interpreted as) the exercise of force prohibited by Article 9 of the Constitution or the deployment of armed forces whose goal would be the exercise of force'.[20] For this reason, the meeting of the so-called 'five PKO principles', including a ceasefire agreement among the combatants and

---

[20] Japan Defence Agency, *Nihon no Boei* (Defence of Japan – Tokyo: National Printing Bureau, 1992), 159-60.

an agreement of the parties involved in the conflict on allowing the Japanese presence, became a prerequisite for Japan's participation.

A second consideration that informed Japanese assessments concerned the impact of concerns upheld by neighbouring countries in East Asia based on their experiences in the Second World War. This point became apparent from the understanding that 'Cambodia has repeatedly expressed the strong hope that the SDF will cooperate with the US Transitional Authority in Cambodia (UNTAC).' In other cases, past memories had less of a weight, and 'some countries have called on Japan to be deliberate in dispatching SDF personnel overseas, in order to contribute to the needs of the international community'.[21] Further, it was necessary to ensure that Japan could choose to cease activities or end the deployment whenever the JSDF was placed under the command of the UN and any of the five PKO principles were no longer being met.[22]

By the mid-1990s, the changes in the international system required Japan to formulate a new defence policy. In the process of drafting what became the NDPO 1995, Prime Minister Hosokawa Morihiro formed an advisory group on defence issues, composed of scholars, former civil servants, and former JSDF officers and chaired by a businessman named Higuchi Hirotaro. The commission issued a report in August 1994 which informed the content of the NDPO 1995. With regard to the stance Japan should take on the participation in international peace cooperation activities, the Higuchi Report stated that '(Japan) should escape from the passive security role it has held until now and behave as an active creator of order'. For this reason, 'It is essential that proactive participation in the various multilateral cooperation efforts carried out within the framework of the UN for international security should, to the extent possible, be seen as an important mission of the SDF.' In the document's view, this commitment carried 'the meaning of providing international public resources for peace'.[23]

This understanding informed the political choices underpinning the NDPO 1995. Alongside the defence of the archipelago and the ability to respond to various situations of major disaster and other contingencies, the NDPO 1995 recommended a clear contribution to the development of a more stable security environment as a main role for

---

[21] Ibid., 161.
[22] Ibid., 160-1.
[23] Advisory Group on Defence Issue, *Higuchi Report*, 13.

the JSDF. From that standpoint, the NDPO 1995 further stressed the promotion of international peace cooperation activities, security dialogues, defence exchanges, and cooperation in the areas of arms control and arms reduction.[24] The NDPO 1976 had focused exclusively on preventive measures against the invasion of Japan and the disruption of the maritime space around the archipelago. By comparison, the NDPO 1995, stipulating the participation in international activities as a new role for Japan's military power, carried a great significance.

The Emergence of New Threats and the Redefinition of the Japan-US Alliance

Japan's contribution to international security was not the only concern of policy planners. Another core concern of Japan's post-Cold War security regarded the growing tensions in Northeast Asia of the early 1990s. This particular aspect became a focal point during the process of redefinition of the Japan-US alliance. The results of this process were the *Japan-US Joint Declaration on Security: Alliance for the 21st Century* by President William Clinton and Prime Minister Hashimoto Ryutaro in April 1996, and the new *Guidelines for Japan-US Defence Cooperation* revised in September 1997. A series of events, including North Korea's declaration to withdraw from the Non-Proliferation Treaty in March 1993 followed by the test launch of a *No-Dong* missile in May 1993, and China's missile fire exercises in March 1996 in an effort to influence Taiwan's first presidential election, played an important role in the redefinition of the security arrangements between Tokyo and Washington.

As mentioned earlier, the Higuchi Report had a significant influence over Japan's post-Cold War security policy. The report called for an active and constructive security policy, noting that 'Japan should extricate itself from its security policy of the past that was, if anything, passive, and henceforth play an active role in shaping a new order.' On the other hand, security experts in Washington were concerned that this stronger activism might come at the expense of the alliance. Indeed, the report recommended that Japan should enhance the functions of

---

[24] Japan Defence Agency, *The National Defence Programme Outline for FY1996 and Beyond* (Tokyo, 1995), http://www.mofa.go.jp/policy/security/defense96/capability .html (Accessed on 10 February 2011). The entire text is available at http://www.mofa .go.jp/policy/security/defense96/.

the alliance, but prioritized multilateral security cooperation.[25] For longstanding Japan observers Patrick Cronin and Michael Green this 'momentum and energy in Japanese policy planning are flowing away from the alliance', and concluded that 'decisive action is now necessary to redefine the alliance.'[26] They further recommended that the Department of Defence (DoD) start 'a comprehensive dialogue with Japan on new bilateral roles and missions.'[27] Based on such recommendations, the two countries engaged in intensive policy discussions to clarify the significance of the alliance.

American officials took the first steps of this bilateral policy coordination. In 1995, Joseph Nye, then Assistant Secretary of Defence for International Affairs, issued *The United States Security Strategy for the East Asia-Pacific Region* widely known as the East Asia Strategy Report (EASR), or Nye Report. The EASR emphasized Asia's new significance, noting that 'United States trade with the Asia-Pacific region in 1993 totalled over $374 billion and accounted for 2.8 million US jobs.'[28] It also reconfirmed American military commitment to the region with 'a force structure that requires approximately 100,000 personnel.'[29] The significance of the alliance between Japan and the US was specifically mentioned as follows:

> There is no more important bilateral relationship than the one we have with Japan. It is fundamental to both our Pacific security policy and our global strategic objectives. Our security alliance with Japan is the linchpin of United States security policy in Asia. It is seen not just by the United States and Japan, but throughout the region, as a major factor for securing stability in Asia.[30]

At the time when this document started circulating in Japan, policy planners were at the final stages of revision of the NDPO first adopted in

---

[25] The report indeed pointed out three core policies in an order as: 1) promotion of multilateral security cooperation on a global and regional scale, 2) enhancement of the functions of the Japan-US security relationship; and 3) possession of a highly reliable and efficient defence capability based on a strengthened information capability and a prompt crisis-management capability.

[26] Patrick M. Cronin and Michael J. Green, 'Redefining the U.S.-Japan Alliance: Tokyo's National Defense Program', *McNair Paper 31*, (Washington, D.C.: National Defense University, November 1994), 2-3.

[27] Ibid., 15.

[28] Office of the Secretary of Defense, *United States Security Strategy for the East Asia-Pacific Region* (Washington D.C.: Department of Defense, February 1995).

[29] US commitment of 100,000 personnel along with the same number for Europe was first declared by the 'Report on the Bottom-Up Review' issued by Secretary of Defense Les Aspin in October 1993.

[30] Office of the Secretary of Defense, EASR, 10.

1976. In November 1995, the National Security Council of Japan and the cabinet announced the new NDPO. The final document sought to provide an answer to the concerns raised by influential American analysts like Cronin and Green and to reflect on their call for a stronger alliance on both sides of the Pacific. It clearly stated that 'The security arrangements with the United States are indispensable to Japan's security and will also continue to play a key role in achieving peace and stability in the surrounding region of Japan and establishing a more stable security environment.'[31] The NDPO further elaborated a newly-introduced dimension of the alliance that is, in short, Japan's role for regional contingencies beyond Japanese territorial borders. In fact, the document referred to a country direct involvement in case a situation was to:

> (…) Arise in the areas surrounding Japan, which will have an important influence on national peace and security, take appropriate response in accordance with the Constitution and relevant laws and regulations, for example, by properly supporting UN activities when needed, and by ensuring the smooth and effective implementation of the Japan-US Security Arrangements.[32]

This point on how Japan should deal with contingencies in the surrounding region, and to what extent Japan was to consider a specific geographic area as 'surrounding the archipelago' stood at the heart of the reconfiguration of the alliance. Part of the answer to these questions arrived in April 1996, when Prime Minister Hashimoto and President Clinton issued the *Japan-U.S. Joint Declaration on Security: Alliance for the 21st Century* in Tokyo. In the declaration, a preliminary consideration was made clear:

> For more than a year, the two governments conducted an intensive review of the evolving political and security environment of the Asia-Pacific region and of various aspects of the Japan-U.S. security relationship. On the basis of this review, the Prime Minister and the President reaffirmed their commitment to the profound common values that guide our national policies: the maintenance of freedom, the pursuit of democracy, and respect for human rights. They agreed that the foundation for our cooperation remain firm, and that this partnership will remain vital in the 21st century.[33]

---

[31] Japan Defence Agency, *The National Defense Program Outline for FY1996 and Beyond*, http://www.mofa.go.jp/policy/security/defense96/capability.html (Accessed on 10 February 2011).

[32] Ibid.

[33] *Japan-U.S. Joint Declaration on Security: Alliance for the 21st Century*, 17 April 1996. Full text available at http://www.mofa.go.jp/region/n-america/us/security/security.html (Accessed on 14 November 2010).

The rest of the joint declaration had three significant meanings. First, as mentioned above, political leaders of the two countries endorsed the redefinition of the alliance so that the Japan-US cooperation in security areas would become vital in the post-Cold War environment. The Prime Minister of Japan and the President of the United States formally declared this to both domestic and overseas audiences. Second, the declaration set a widened agenda for the two countries by mentioning three areas for cooperation: (a) bilateral cooperation under the Japan-US security relationship, (b) regional cooperation, and (c) global cooperation. During the Cold War period, the alliance was narrowly defined and focused only on the bilateral context, specifically referring to the US commitment to the defence of Japan, Japan's support of American forces stationed in Japan, and bilateral cooperation in military technologies and platforms. The two leaders made clear that Tokyo and Washington would cooperate on regional and global issues as well. For example, joint efforts to improve the stability of the Korean Peninsula and enhance regional security dialogues are listed as key areas for regional cooperation. Strengthened cooperation in support of international peacekeeping and humanitarian relief operations, and coordination between the two countries for arms control and disarmament are included in global cooperation. The scope of the bilateral cooperation became wider than that of the Cold War period. The third, but not least important aspect was the proposal for revision of the Guidelines for Defence Cooperation.

On this basis, the two governments started efforts to revise the *Guidelines for Japan-US Defence Cooperation*. The first Defence Guidelines had been authorized by the Japan-US Security Consultative Committee (SCC) and adopted by the National Defence Council and the Cabinet of Japan in 1978.[34] This first document was the result of two years of continuous work by officials from the two countries committed to a smoother and more effective implementation of the bilateral military partnership, begun in 1976 under the aegis of JDA Secretary General Sakata Michita and Defence Secretary James Schlesinger. Whilst the first Defence Guidelines played an important role to facilitate enhancement of operational cooperation between the

---

[34] The Security Consultative Committee was established in 1960. The SCC included the top representatives of the Ministry of Foreign Affairs and the JDA from the Japanese side, the US Ambassador to Japan, and the Commander of the Pacific Command for the United States. In 1990, the level of US membership was raised to Secretaries of Defense and States.

two militaries through the late 1970s and the 1980s, especially in the maritime realm, they were designed to deal with a Cold War strategic environment.[35] As Japan and the US entered a post-Cold War environment, it was inevitable to review the guidelines. In August 1996, government officials and military planners from the two countries started working on a set of measures to take the military partnership into the twenty-first century. After intensive discussions, including four case studies on various contingency scenarios, the revised Defence Guidelines were reported to the SSC in September 1997 in Washington.[36]

The revised Defence Guidelines consisted of three major elements: (a) cooperation under normal circumstances, (b) actions in response to an armed attack against Japan, and (c) cooperation for situations in areas surrounding Japan that will have an important influence on Japan's peace and security (Situations in Areas Surrounding Japan). In short, the revised guidelines addressed bilateral military cooperation across the full spectrum of Japan's national security items, in peacetime and in case of crisis, in cases of attacks to the archipelago as well as in regional contingencies. Peacetime cooperation included coordination for activities such as peacekeeping operations, humanitarian relief operations, security dialogues and defence exchanges as well as information sharing, policy coordination and integrated defence planning between the two governments. For the second element, Japan's defence, the guidelines made clear that 'Japan will have primary responsibility to immediately take action and to repel an armed attack', whilst American forces 'will provide appropriate support varying according to the scale, type, phase, and other factors of the armed attack'.[37]

Issues related to the third element, cooperation for situations in areas surrounding Japan created an intense debate among lawmakers and national and international observers. One key aspect concerned the geographic area of application which remained only vaguely defined. The revised Defence Guidelines focused instead on items of cooperation and divided them into three categories. The first category concerned cooperation in activities initiated by either government

---

[35] Tanaka, *Anzen Hosho*, 382-4.

[36] Masaru Honda, 'Reisengo no Doumei Hyoryu' (Drifting of the Alliance after the Cold War), in Sotooka, Honda and Miura, *Nichibei-doumei Hanseiki*.

[37] The document *The Guidelines for Japan-US Defence Cooperation* was submitted by the Subcommittee for Defence Cooperation and authorized by the Security Consultative Committee on 23 September 1997.

such as relief activities, measures to deal with refugees, search and rescue operations, non-combatant evacuation operations, and activities for international sanctions including UN-endorsed ship inspections. The second category specified Japan's support to US forces for the use of facilities in Japan and for logistics. The third category included operational cooperation such as Japan's activities for intelligence gathering, surveillance and minesweeping operations, and American operations to restore the peace and security in the areas surrounding Japan, with an implicit understanding of the immediate regional neighbourhood in Northeast Asia. Forty specific items for cooperation were listed in total in the appendix of the revised Defence Guidelines.

The Guidelines were not a legally binding agreement and therefore, 'the guidelines and programs under the guidelines will not obligate either government to take legislative, budgetary or administrative measures'. Nonetheless, the two governments were 'expected to reflect in an appropriate way the results of these efforts, based on their own judgment, in their specific policies and measures'.[38] As such, the Japanese government started taking a series of measures to implement the content of the new guidelines. On 24 September 1997, the day after the Defence Guidelines were reported, Japan adopted a Cabinet Decision to initiate work to ensure their effectiveness. It was made clear that the work should be done by the government as a whole rather than only the JDA or the JSDF to address a range of possible legal issues. By April 1998, the administration had submitted bills related to Japan's action in case of 'Situations in Areas Surrounding Japan' to the Diet. In May 1999, the Diet passed the *Law Concerning Measures to Enhance the Peace and Security of Japan in Situations in Areas Surrounding Japan*. In the meantime, minor amendments were made on the JSDF *Law and the Acquisition and Cross-Servicing Agreement between Japan and the United States* (ACSA) to support the new normative measures. During the process of deliberation on these bills, the Diet decided to take separate measures for ship inspection operations. The bill concerning ship inspection operations was submitted to the Diet in October 2000, and enacted in the following month.

---

[38] *The Guidelines for Japan-US Defence Cooperation*, 23 September 1997, http://www.mofa.go.jp/region/n-america/us/security/guideline2.html (Accessed on 20 November 2010).

This series of legislative actions enabled Japan to cooperate with US forces in the widened scope of the alliance and relevant governmental agencies as well as the JDA and the JSDF were tasked to implement measures according to respective jurisdictions. Heads of such agencies were authorised to require cooperation of local governments, and to request cooperation from non-governmental bodies. This legislation set several conditions to avoid political and constitutional controversy over the activities of Japan including the JSDF. First, these activities had not to lead to the threat or the use of force. Second, these activities had to be conducted in areas where there is currently no combat action and where it is deemed that no combat action will take place during the term of the activities. Third, the Prime Minister had to obtain prior Diet approval for the deployment of the JSDF except in cases where implementation of such activities was deemed urgent.[39]

This process highlighted a key element in the interpretation of the notion of 'areas surrounding Japan'. The scope of Japan's contribution was to be more 'situational' than 'geographic' and pre-eminently related to the functions and overall context within which the JSDF would operate. This was in line with the spirit of the post-war Constitution, which focused on the question of the use of force. Former administrative vice minister for defence, Masahiro Akiyama, explained this point, stressing that 'what Japan can do and what Japan cannot do under the Constitution were made clear', and that 'the laws enabled Japan to offer as much cooperation as possible for the United States under the conditions that did not allow the exercise of the right of collective defence'.[40] This interpretation later facilitated Tokyo's efforts to enact legislations that enabled the dispatch of the JSDF in the Indian Ocean, under the *Anti-Terrorism Special Measures Law* of 2001, and in Iraq, under the *Special Measures Law for Reconstruction and Humanitarian Assistance in Iraq* in 2003. In the drafting of these bills, the Japanese government no longer had to verify the constitutionality of the activities as long as they were included in the laws concerning the situations in areas surrounding Japan.

---

[39] Japan's defence White Papers of recent years contained detailed explanation of the laws. See, for example, Japan Defence Agency, *Defence of Japan 2002* (Tokyo: Urban Connection, 2002), 189-93.

[40] Masahiro Akiyama, *Nichibei no Senryaku-taiwa ga Hajimatta* (*The Japan-U.S. Strategic Dialogue* – Tokyo: Aki-Shobo, 2002), 270.

By the end of the 1990s, a renewed series of direct regional threats further stressed the significance for Japan to strengthen the content of the guidelines, focusing on bilateral cooperation to meet the challenges presented by conventional and nuclear ballistic missiles and guerrilla-commando type attacks.[41] North Korea's *Taepo-Dong* missile launch in 1998 made the threat of medium range ballistic missiles apparent. In March 1999, two spy ships sent by North Korea violated Japanese territorial waters, and the JMSDF was deployed for the first time in its history in support of the Japanese Coast Guard to pursue those ships. Shortly after, as the issue of the abduction of Japanese nationals by North Korea became publicly known, Japan found itself highly exposed to irregular threats.

The framework of the guidelines proved to be useful to complement Japan's own initiatives in order to pursue adequate responses to the increase set of security threats. Notably, in 1999, the Japanese government decided to begin a joint technical research on a *Sea-Based Mid-course Defence System* to counter attempts of missile attacks. Concurrently, the JASDF sought to upgrade its air defence missile forces by acquiring the *Patriot Advance Capability-3* (PAC-3), eventually deploying the first batteries in 2007. In order to address lower intensity threats like the special force commandos, the JMSDF created a Special Boarding Unit in March 2001. This was followed by the activation of JGSDF Special Operation Group in 2004.

Conclusions

This chapter has discussed changes in perception of security threats and of strategic priorities of Japan's defence policy from the end of the Second World War throughout the Cold War and in the first decade of post-Cold War periods. Changes in the security environment over the last decade of the twentieth century triggered an evolution in defence affairs that required Japan to reach beyond the borders of the archipelago. As the century drew to a close, one issue still remained to be resolved, namely how to strike the right balance between the requirements for the defence of the Japanese archipelago and the participation in international peace activities.

---

[41] Ibid.

This entailed two different orders of problems. The first concerns the adequacy of the legal provisions authorizing the JSDF operations dealing with direct attacks to the archipelago. The debate over post-Cold War defence policy began with a focus on overseas operations of the JSDF then shifted to regional contingencies outside of Japan. Meanwhile, the legal basis for JSDF operations in the defence of Japan remained untouched. In 1976, a Soviet MIG-25 fighter landed into Hakodate airport, in southern Hokkaido, after violating Japanese air space. The inability of the JSDF to deal with such an emergency due to lack of proper legal arrangements, prompted the government to explore more suitable norms for action. Only in 2001 a series of laws were enacted to ensure that the JSDF could properly intervene to defend Japan. Provided the changing nature of the direct threats to Japan by the end of the 1990s, tests in the form of tabletop and field exercises in a variety of scenarios involving relevant government and non-government organizations should be introduced to verify the adequacy of the current solutions.

The second set of issues was related to JSDF's adaptability to international operations. The force structure of the JSDF was originally designed to protect the homeland utilizing existing domestic infrastructures including commercial transportation, medical and distribution systems. This force structure was later found insufficient to support JSDF troops deployed abroad. Since 1992, when Japan dispatched UN Peacekeepers to Cambodia for the first time, and whenever the JSDF dispatched contingents for international missions, government authorities had to refer to 'special units' due to the lack of organizational support capabilities. The command structure of the JSDF also had similar problems. For example, since the JGSDF was organized into five regional armies in charge of defending their respective areas of responsibility without any organizations responsible for missions abroad. One of these regional armies' headquarters was assigned to plan and execute each international peace operation. Without expertise in international operations, commanders and staff officers had to carry out their tasks without any past experience, thus the readiness remained relatively low. This problem was properly addressed only in the 2000s, with the activation of the Central Readiness Force in 2007, whose headquarters' primary tasks included planning and execution of all the JGSDF overseas operations.

The balance between resources allocated to those two areas will remain critical in the future along with the serious question of how much the country will be able to afford both financially and in terms of personnel. Even for the defence of Japan alone, a key question remains on the balance between the traditional capabilities for deterring and responding to the use of conventional military means, and those for dealing with various non-traditional threats such as the use of chemical and biological weapons by non-state actors, or other illegal activities including the trafficking of drugs or weapons of mass destruction-related materials. This question of balance between forces for regional and homeland defence and capabilities for expeditionary missions has become the hallmark of Japan's military dilemma at the beginning of the twenty-first century.

CHAPTER FIVE

BRITISH DEFENCE POLICY AND THE TRANSFORMATION OF THE ROYAL NAVY IN THE COLD WAR AND BEYOND[1]

Eric Grove

Never had British power been so widely dispersed as in 1945, and never had its foundations been weaker. The Attlee administration had the enormous problem of reviving an economy bankrupted by war, constructing a new, more equitable domestic social order, and deciding on an appropriate level of military strength. Ambitious plans for the post-war navy, as well as for the rest of the armed forces, remained dormant. In the Cabinet's initial discussions on Britain's post-war defence posture, the Prime Minister conceded American supremacy on the seas. Dogged attempts by A. V. Alexander, the First Lord of the Admiralty, to put strategic priorities rather than resource allocations as the main parameters in the defence policy equation, fell on stony ground. In Ernest Bevin's homely metaphor, 'We must cut our coat according to our cloth.'[2]

There was no question of completing wartime construction plans. This was evident by the impact of the early post-war defence reviews on the fleet's main punch, its aircraft carriers. Only two fleet carriers, *Eagle* and *Ark Royal* and four light fleet carriers, *Albion*, *Bulwark*, *Centaur* and *Hermes* survived in the yards, to be completed as soon as practicable. This was a minimum force capable of operating the heavier and faster aircraft types planned for the future. The existing carrier fleet made up of armoured hangar carriers of pre-war conception and the still incomplete programme of light fleet carriers built for Second World War aircraft had only limited development potential. The six *Majestic* class light fleet carriers still under construction were suspended and none saw service with the RN. The immediate operational post-war

---

[1] For fuller accounts, see Eric Grove, *Vanguard to Trident: British Naval Policy Since World War II* (Annapolis, MD: Naval Institute Press, 1987); and E. Grove, *The Royal Navy Since 1815* (Basingstoke: Palgrave Mamillan, 2005).

[2] UKNA, CAB69/7: Minutes of the 6th and 11th Meetings of the Defence Committee (Operations).

carrier force was provided by their near sisters, six affordable *Colossus* class light fleets.

This chapter addresses the naval dimension of British post-war defence policy from the initial years of the Cold War to the end of the twentieth century. Post-war Britain was different from Japan – it was emerging victorious from the war. Yet, the war had had immense costs and had irreversibly damaged the Empire and its economic foundations. In post-war Britain, the maintenance of a degree of sea power centred on the core issue of how to strike the balance for a fleet that could act as the country's military and diplomatic spearhead on the global stage, as well as its first line of defence in European waters. Britain had more than enough ships left over from the war: the problems were deciding which to continue running in the operational fleet, how many to place in reserve, and how many to dispose of. Those vessels that were too old and worn-out were soon scrapped. Many others were decommissioned and filled British dockyards and harbours, poised to be mobilised in any future, traditional conflict. But, if these ships were to be used in the future, they required refit and maintenance facilities – a drain on the active fleet. The chapter examines how across more than five decades British defence policy strove to provide an answer to the above questions and how, at all times, sea power remained a vital tool of statecraft.

A 'Restricted Fleet' for the Early Post-War Period

In February 1946 it was decided by the Cabinet's Defence Committee that there would be no war in the next two or three years and that no hostile fleet would exist for the next few years. No naval building programme was drawn up for 1946 and planned naval estimates and manpower ceilings were reduced. The Prime Minister remained dissatisfied. Why, he asked in July 1946, were 182,000 male naval personnel required in 1947 when 119,000 had been sufficient to man a much larger fleet in 1938? The Admiralty's reply shows the different requirements of the more complex post-war Navy.

> The introduction of the large carrier, the heavy increase in anti-aircraft guns, the immense growth in technical equipment, the development of radar and W/T, the speed of the modern commerce destroying submarine, the advent of combined operations and the introduction of short term conscription have substantially changed the situation and their cumulative effect has been largely to increase the number of men

required to maintain a given number of naval ships, units and training depots and schools.[3]

The year 1947 began quite well for the navy, with attempts to formulate coherent plans for the development of post-war armed forces. The 'Defence of sea communications' was a priority in British defence policy, second only to the 'Defence of the United Kingdom base'. In April 1947, a ten-year plan provided a time frame for the creation of forces ready for war. There was a 'small' risk of war over the next five years and a gradually increasing one for the five years after that. Yet, the Admiralty's first plans were totally unrealistic: a peacetime active fleet of 128 major surface ships, including three battleships and four fleet carriers, twenty-nine submarines and 500 front-line aircraft. The wartime fleet would comprise over 600 major units and over 1,000 aircraft. A £3.3 billion programme was produced which would cost some £465 million annually by the mid-1950s. In the prevailing atmosphere of economic crisis, Alexander told the three services to budget on a *total* allocation £600 million. The 1957 assumption became very close to being a 'five-year no war rule', albeit one fixed not rolling, as the pre-war ten-year rule had been. The Admiralty planners set about revising their plans downwards.

By the end of 1947, the RN was in a parlous state. The Home Fleet, with many of its units 'temporarily immobilized', was composed of a tiny operational cadre of a cruiser, a couple of large destroyers, half a dozen frigates and twenty submarines. The Mediterranean had the only 'fleet' worthy of the name, with one operational light fleet carrier with the new standard air group of twenty-eight aircraft, a squadron of *Seafires* and a squadron of *Fireflies* (US aircraft had been discarded with the end of lend lease as the dollar situation precluded acquiring spares). Also in the Mediterranean were four cruisers, twenty destroyers and frigates and two submarines. The Pacific Fleet had temporarily lost its carrier and was down to three operational cruisers, four destroyers, four frigates and three submarines. A cruiser and two frigates were operational in the South Atlantic and a similar force was on the American and West Indies Station. Two cruisers and two frigates composed the East Indies Squadron at Trincomalee and a lone frigate kept the White Ensign flying in the Persian Gulf.

---

[3] UKNA, CAB131/3, Former DO(46)97: Defence Committee, Minutes and Papers, 30 May, 31 December 1946.

The Naval Staff finally drew up a practical 'Revised Restricted Fleet' plan in the middle of 1949. This was to be capable of carrying out the foreign and colonial policy of the government in peacetime, and then to be able to meet the immediate requirements of a 1957 war, serving as a nucleus for expansion. It contained no battleship in full peacetime commission. The warship HMS *Vanguard*, incomplete until August 1946, would be retained in service for training and royal yacht duties; the four surviving units of the *King George V* class would be added to the Reserve Fleet. Two fleet carriers and three light fleet carriers would be kept active, with one of each in addition for training. There would be 250 front-line carrier based aircraft. Thirteen cruisers would be in peacetime commission, thirty-eight destroyers, thirty-two frigates (the term now used to cover corvettes, escort destroyers and sloops as well as the old frigates proper) and twenty submarines. A large reserve of small minesweepers would be built up and over a hundred frigates would be kept in reserve. Some fifty fast patrol boats (as MTBs and MGBs were now known) would be kept for training and in reserve. Such a fleet could maintain British peacetime interests worldwide and, with the USN, take half the wartime responsibility for the Atlantic and Mediterranean.

The 'Revised Restricted Fleet' still emphasized 'hot war' priorities in new construction. The threat of mining, a type of warfare for which the Soviet Union had known predilection, was reflected in the plans for large numbers of new construction coastal and inshore mine counter-measures (MCM) vessels. Other new construction investment was to go into frigates for convoy escort. Concurrently, wartime fleet destroyers were to be converted into fast *Type 15* and *16* ASW frigates to provide a readily available answer to the menace of the much faster modern submarine. The heavy backup to these smaller vessels would come from the two new fleet and four new light fleet carriers, supplemented by extensive modernization to some of the older fleet carriers. One fleet carrier, HMS *Implacable*, was returned to full commission with the Home Fleet. She had a special air group of twin-engined *Sea Hornet* fighters and *Firebrand* fighter-torpedo bombers.

The Admiralty had learned the lessons of the recent war. With large surface ships known to be under construction in the Soviet Union it could not abandon the 'essential insurance' of battleships and other powerful surface warfare vessels; new large *Daring* class destroyers that had survived the 1945 cuts were continued and a still larger 5,000 ton cross between destroyer and cruiser were planned. Nevertheless, the three-dimensional nature of naval warfare was clearly recognized. The new First Sea Lord, Lord Fraser of North Cape, who took up office in 1948, put the direct protection of shipping from underwater and air threats as the top wartime priority: 'Planning can only proceed on something we know we must do. Escort safely our convoys.'[4] Fleet carriers would cover convoys in the Mediterranean with their fighters; light fleets would act as escort carriers with fighters and ASW aircraft in the Atlantic.

The Cold War Goes 'Warm': Projecting Power in Korea

Lessons drawn from the last war were put to a different test when limited war broke out in Korea in 1950, and the light fleet carriers were used primarily against land targets. The carrier HMS *Triumph* operated her *Seafires* and *Fireflies* at the outset of the conflict and her sisters kept a Commonwealth carrier on station throughout the conflict. The RN played a significant role in this conflict, being responsible for the blockade of the western coast of the Peninsula for most of the war. British cruisers played a major part in covering MacArthur's landing at Inchon and, together with destroyers and frigates, kept up a constant bombardment on the Chinese and North Koreans, firing over 170,000 rounds of ammunition to add to 23,000 carrier sorties. The war not only provided a useful demonstration of the RN in 'warm' war as well as 'cold', it also provided the context for the last post-war attempt to build up a fleet to fight an old-style war.

The Korean War launched Britain into a programme of rapid rearmament. The 1957 Planning Date was increasingly replaced by the North Atlantic Treaty Organization's date of 1954. The force goals set out in NATO's Medium Term Plan provided an ambitious target for a rapid naval build-up. Nevertheless, the 'Fraser Plan' of October 1950 showed only modest increases over the Revised Restricted Fleet, with an expanded frigate construction and conversion programme and even more mine countermeasures vessels. In December, after Attlee's talks with Truman in Washington, Britain began to rearm to the limit of her capability. An 'Accelerated Fraser Plan' was produced, with an emphasis on rapid results, even more MCM vessels, an accelerated frigate

---

[4] UKNA, ADM205/69 First Sea Lord's papers, Correspondence with C's in C Mediterranean and Far East and Commonwealth countries regarding long term policy, 1948.

conversion programme and the rapid procurement of available types of aircraft. The new plan would cost £1,610 million, over a third of the entire £4.7 billion revised rearmament plan.

It was becoming all too clear, however, that Britain was not capable of spending the money allocated to naval rearmament. Given the reluctance of the government to go over to a full war economy, the rearmament programme was beset by shortages of labour and materials. Work was seriously delayed by a shortage of drawing office staff, both at the Admiralty's offices and the shipyards. Electricians, plumbers and fitters were also unavailable. Pipes, valves and electrical fittings were delivered late. Labour disputes caused further delay. Moreover, even if the resources now flowing into defence were allowing some of the ships left over from the war to be completed, the navy's chronic manpower shortage, despite increased conscription and the recall of reservists, meant that older ships had to be decommissioned to man the newer ones.

Before the end of 1951 the rearmament programme had been cut back, fatally delaying some projects including a second fleet carrier conversion. The priority remained on short-term projects like the *Daring* class. For the longer term, the government had something more radical in mind. In 1952, Britain exploded her first nuclear bomb. Intelligence assessments seemed to demonstrate that the Soviet Union might well be deterred from starting a war for quite some time. Rearmament to fight a war had to be replaced by an affordable deterrent posture for the long haul. The Chiefs of Staff were tasked with formulating a revised Global Strategy for the new conditions. When their first attempt proved unsatisfactory, the famous week-long meeting took place at the Royal Naval College Greenwich, when the Chiefs formulated a revised Defence Policy and Global Strategy paper, which was forwarded to the Prime Minister in June 1952 and adopted by the Cabinet's Defence Committee the following month.[5]

The main thrust of this paper was that nuclear weapons, especially those delivered by the increasingly powerful US Strategic Air Command, meant that any future war promised to be short and intense. The First Sea Lord, Admiral Sir Rhoderick McGrigor, was not willing to abandon the long war scenario completely – 'alternative loading and unloading facilities' were being investigated to replace the major ports – but he agreed with his colleagues that 'the fact that it is economically

---

[5] It can be found at UKNA, DEFE 5/40 Chiefs of Staff Committee, Memoranda 13 June – 15 August 1952, as COS(52)361.

impracticable to make the preparations necessary for a long war should be faced, and a guiding principle of a rearmament programme should be to ensure survival in the short opening phase'.[6] McGrigor accepted that the fleet ninety days after mobilization in 1955 would now be one battle-ship, five aircraft carriers (280 aircraft), fourteen cruisers, 163 destroyers and frigates, 263 MCM vessels thirty-nine submarines and three fast minelayers, a reduction in the NATO requirement of his was one carrier, four cruisers, 113 destroyers and frigates and 104 MCM vessels.

Even the reduced force levels reluctantly agreed upon at Green-wich proved too extensive for Butler to accept. Throughout 1952 a struggle continued as the Chancellor fought for still further reduc-tions in defence expenditure. A compromise, agreed at the end of the year, involved cuts in manpower that meant that many ships were only semi-operational. Appearances were, however, kept up. External observers were not to know that the impressive British cruiser on the port visit had only 30% of her armament manned or that the mag-nificent HMS *Vanguard*, now returned to notional full commission as Home Fleet flagship, usually had unmanned turrets and magazines and normally no ammunition on board except starshell for her secondary battery. Such solutions maintained British prestige in the short term as demanded by the Foreign Office. *Vanguard*'s role was officially classi-fied as a peacetime one – she was not required for war until some three months after mobilization.

In 1953 a more 'Radical' Review began to bring defence policy into line with Butler's Treasury resources. The key to this reassessment was an extension of the logic of Global Strategy. Only those forces that con-tributed to the United Kingdom's world power status and which were relevant in the first six weeks of war were to be retained. In 1952, the Chiefs of Staff had emphasized the importance of an 'immediate attack at source against airfields, U-boat bases and mine depots' but the RAF and the Minister of Supply, Duncan Sandys, saw this as a role for the RAF's planned medium bombers and mounted a serious joint attack on the Admiralty's fleet carrier and aircraft plans.

This was a serious threat to the RN whose main wartime rationale moved steadily towards the carrier strike role. This reflected closer British involvement with the NATO Strike Fleet concept and also the belated appearance, to the Admiralty's relief, of a surface threat, the *Sverdlov* class cruiser, which seemed to have much potential as a commerce raider.

---

[6] Ibid.

These were only part of the story. Forced to accept a subordinate position in the new NATO Atlantic command structure (the C-in-C Home Fleet was NATO C-in-C Eastern Atlantic), the Admiralty was determined to wield as much influence over the Alliance's main fleet as possible. This implied the retention of a fleet carrier force equipped with high-performance fighters to cover US carriers and, eventually, strike aircraft to add their nuclear weight to the American carrier offensive. In June 1952 a requirement was issued for a high-performance carrier-based NA39 (Buccaneer) nuclear strike aircraft and it was hoped that in the meantime the next generation N113 (Scimitar) and DH110 (Sea Vixen) fighters would provide an interim nuclear capability. The British contribution to the Strike Fleet might have to hold the ring in northern waters alone before the arrival of American ships.

The competition between the Services, and between Sandys (who had charge of aircraft procurement) and the Admiralty, became acute. In 1953 the Naval Staff abandoned its plans for a new 53,000 ton carrier, although it was still planned to lay down a 35,000 ton ship in 1957.[7] In the shorter term it would be lucky to save the one modernization of a wartime carrier that had actually begun, HMS *Victorious*. The Minister of Defence, Lord Alexander of Tunis, suggested a compromise that would have completed HMS *Ark Royal*, to join her already commissioned sister *Eagle*, but which would have only kept one ship in commission, equipped only with fighters and ASW aircraft for escort duties. In their opposition to these proposals, the Admiralty argued that abolishing the fleet carrier would decisively weaken British claims to NATO commands as well as generally diminish the UK's international status.

Fighting the Cold War at Home

A compromise was offered to the RAF in which it was agreed that, whilst bombing shore targets was its primary responsibility, the carriers would be required to attack the *Sverdlov* vessels at sea. The utility of the carrier in Cold War functions also helped save the day for the Admiralty. In mid-1954, a revised Chiefs of Staff Appreciation in the light of the H-bomb made the Cold War relevance of forces the over-riding factor in their retention in Britain's force posture. The Admiralty were

---

[7] D. K. Brown and George Moore, *Rebuilding the Royal Navy: British Warship Design Since 1945* (London: Chatham, 2004), 56.

willing to give up a light carrier from their proposed active force, reducing the latter to two fleets and two light fleets. This was in line with the decreed de-emphasizing of escort duties inherent in the revised strategic concept. In a subsequent review, however, budgetary constraints resulted in decisions that went against the Admiralty's wishes on the carrier question. It concluded that:

> In the new strategic conditions, the relative importance of sea power in our defences is evidently diminishing and there is no sign that the trend will be arrested. There is no question of having a larger navy than we need or can afford; and we must make the best use of existing material. It is natural that the Navy should wish to have their (sic) share in air power, which is growing in importance. The cost of the Fleet Air Arm, however, . . . appears to impose a burden disproportionate to the results. Moreover, the role of the aircraft carrier is already restricted through the ever-increasing range of shore-based aircraft.[8]

The new review recommended that all four carriers should be manned and equipped as escort carriers only, with the primary role of protecting shipping in the later stages of a future war. The Admiralty reacted with a strongly-worded protest, signed by the First Lord. McGrigor led the argument in Cabinet stressing the need to have a British contribution to the main NATO Atlantic Striking Fleet. He reiterated the importance of the British ships before the Americans arrived and the protection of Norway from amphibious attack. Churchill, very reluctantly, accepted that alternative naval economies might be made and tasked Macmillan, his new Defence Minister, to continue negotiations with the Admiralty on alternative cuts.

These were successful, for when the Defence White Paper and Naval Estimates appeared in February 1955, the continued presence of the carriers as 'the fists of the fleet' and 'the strength upon which all naval activities depend' was strongly asserted. The Allied striking fleet's carriers, it was even argued, 'will add powerfully to our ability to hit the enemy either independently or in support of allied land forces and land-based air forces'.[9] It was apt that *Ark Royal* should now appear to join *Eagle* and the two light fleet carriers *Centaur* and *Albion* in the first line carrier force. They were beginning to operate the first generation of jet aircraft, the *Sea Hawk* single-seat fighter and *Sea Venom* two-seat all-weather fighter. Following the failure of the last variant of the *Firefly*

---

[8] UKNA, COS(54)250.
[9] Command Papers (Hereafter Cm) 9396, paragraph 12, Statement on Naval Estimates 1955.

in the anti-submarine role, American *Avengers* had filled the gap before the turboprop *Gannet* entered service. Much, however, had been forfeited. The minesweeper programme was drastically reduced despite the Admiralty's protests that the smaller ports that had not suffered nuclear attack might still face a mining threat. Nonetheless, a large number of 'Ton' class coastal minesweepers, 'Ham' class inshore minesweepers and some 'Ley' class inshore minehunters were built. Those proved useful for fishery protection and patrol duties.

Frigate building plans were reduced. The cruiser fleet was to be cut and its modernization drastically curtailed as sophisticated anti-aircraft armaments were now deemed to be unnecessary against unsophisticated Cold War opposition. The Admiralty still planned to have a few cruisers equipped with modern AA systems, notably the three incomplete *Tiger* class ships, upon which work began once more after a long sojourn in Scottish lochs. It also planned to build a totally new large missile cruiser equipped with the large *Seaslug* missile system, upon which work had been going on for a decade. New fleet escort vessels were planned and the *Type 12* anti-submarine frigates had enough speed to be useful carrier task force escorts. What specialized convoy escorts had been built, however, would still be useful to escort replenishment groups and act as gunboats. The long review process had seen some notable casualties, among them the 5,000-ton cruiser/destroyer, an east coast gunboat to protect convoys from attack by coastal forces, an even more austere mass-production ASW frigate, and a new ocean minesweeper.

It could have been much worse for the Admiralty. Churchill had 'lost interest in the Navy' and his resignation in 1955 was something of an advantage for its interests. He did not want a strong First Sea Lord and had unsuccessfully opposed the appointment of the influential Lord Louis Mountbatten to replace McGrigor in April 1955. The old Prime Minister was quoted as saying he did not want a strong man in charge of it. Mountbatten was an unsurpassed, and often unscrupulous bureaucratic politician. Moreover, his social status and connections, which had sustained him through an equivocal record as a commander, now stood the Service in excellent stead to weather the storms.

Mountbatten's arrival at the Admiralty more or less coincided with Anthony Eden's at Downing Street, and the First Sea Lord exerted as much influence as possible on the new Prime Minister. Eden was determined to emphasize cuts in general war forces in future defence reviews and in 1956 set up a Policy Review Committee to draw up plans for forces based on the assumption that the UK would be knocked out in a thermonuclear war. The navy should be reduced to 'the minimum necessary for situations short of global war'.[10] The Admiralty still rejected this logic but it was forced to accept that global war forces were the lowest priority and that its future was best safeguarded by an emphasis on cold and limited war. Indeed this would require additions to the active fleet, including a new 'commando carrier', a light fleet carrier converted to land Royal Marines by helicopter (plans for the new strike carrier were quietly dropped).

In the second half of 1956 the weaknesses of Britain's existing limited war forces were shown up dramatically. When President Nasser nationalized the Suez Canal immediate action was impossible and when the attack eventually went in to recover the Canal and topple Nasser, the political context was all wrong and failure inevitable. The operation allowed the navy to make various useful points. It proved the greater effectiveness of carrier air power for air support, of the commando carrier concept, of the continued importance of sealift and amphibious assault, and the crucial importance of rapidly deployable forces which could act when the political context was right. Suez also implied that those forces need not be too large as the capitulation to American pressure demonstrated that the kinds of action that Britain would undertake alone would be relatively limited.

The departure of Eden after the humiliation of Suez led to the culmination of the long process of Conservative Defence Review. The new Prime Minister, Macmillan, appointed the navy's old enemy, Duncan Sandys, as Minister of Defence, vested with far-reaching powers for root-and-branch reform, notably the abolition of conscription. Sandys and Mountbatten fought out what the latter called a 'pretty good tussle' in the early months of 1957. Much to the Defence Minister's surprise, complete Chiefs of Staff endorsement of the carrier was obtained. It was agreed that the Navy's role in general war was 'somewhat uncertain' but the White Paper confirmed the continued utility of Naval forces as a 'means of bringing power rapidly to bear in peacetime emergencies or limited hostilities'.[11] Although the RN at home was to be reduced, forces East of Suez were to remain as a carrier-centred 'fire brigade'

---

[10] Quoted in Martin S. Navias, *Nuclear Weapons and British Strategic Planning 1955-58* (Oxford: Clarendon Press, 1991), 75.

[11] Cm 124, Defence, Outline of Future Policy.

(in the homely metaphor of the Admiralty's own Statement on the Naval Estimates).[12]

One historic decision was to establish that traditional cruiser roles could be taken over by carriers and 6,000-ton guided missile destroyers, the design of which was presented to the Board in March 1957. The following month it accepted the abandonment of the 16,000-ton guided missile cruiser in which much work and publicity had been invested 'if the future of the GW (guided missiles, ndr.) destroyers and the four carriers were assured'.[13] The force of nine cruisers in commission and four in reserve was to be run down to the three *Tigers* which were to be completed as 'A great deal of money has been invested in these ships, and apart from awkward criticism if they were again suspended or cancelled, their ultra modern gunnery systems will make them rather better than stop gaps until the new destroyers come into service in adequate numbers.'[14]

Sandys still did not believe in the Striking Fleet concept. As the second phase of his Review went on in 1957, the Admiralty emphasized that the USA was committed to the Striking Fleet and that it would be better for Alliance relations if Britain continued to make a contribution, just as she did to NATO's forces in Europe. The Ministry of Defence was persuaded that, even if no provision was made for fighting a global thermonuclear war, the Navy had an important role in meeting Soviet challenges that might not necessarily bring the deterrent into play. Naval forces also had a vital role in deterring limited war and fighting it if it occurred. It also had an 'unquestioned' role in 'imperial policing'.[15] On this basis Mountbatten won the argument on retaining a strength of 88,000 personnel; he considered he had got the Navy a 'reasonable deal – better than the Army and RAF'.[16]

He had to make a few compromises. He agreed that the two carriers west of Suez should have anti-submarine air groups, but even this concession was short-lived. Because of operational shortcomings it was decided to hasten the withdrawal of the existing ASW aircraft, the *Fairey Gannet*. It was replaced between 1958 and 1960 by the

---

[12] Cm 151.

[13] UKNA, ADM 167/150 Board of Admiralty: Minutes and Memoranda, 1957.

[14] Ibid.

[15] UKNA, DEFE 7/1669 Ministry of Defence Role of the Navy Paper, 27 September 1957.

[16] Philip Ziegler: *Mountbatten: The Official Biography* (London: Harper Collins, 1985), 553-4.

Whirlwind helicopter. By the time this was done, Harold Watkinson was Minister of Defence and he allowed all four operational carriers to retain mixed, general-purpose air groups. From 1961 these included an improved *Wessex* ASW helicopter. The *Gannet* continued in a new form as an airborne early warning aircraft carrying the radar formerly fitted to the *Skyraiders*.

A New Cold War Navy

A new Navy was taking shape. The battleships disappeared, HMS *Vanguard* finally going for scrap in 1960. The last of the *Hermes* class light fleet carriers (*Hermes* herself) was completed in 1959 with a fully angled deck and steam catapults to operate the latest aircraft. The carrier *Victorious*, the only wartime veteran to receive a full rebuild, also finally joined the fleet, similarly equipped. Both also carried the latest 984 three-dimensional radar combined with a pioneering comprehensive display system analogue action information system. They carried high-performance swept-wing aircraft, the single-seat *Scimitar* and two-seat all-weather *Sea Vixen* for which *Red Beard* 15-kiloton nuclear weapons became available in 1961. The longer ranged *Buccaneer* strike aircraft replaced the *Scimitar* from 1962. In the mid-1960s, two existing light fleet carriers were converted into commando ships with *Whirlwind* (later *Wessex*) transport helicopters, and two modern assault ships finally ordered. The Royal Marines were concentrated into an expanded Commando force.

The first of ten planned *County* class guided missile destroyers was begun, combining four 4.5-inch guns with *Seaslug* and *Seacat* anti-air missiles and a *Wessex* ASW helicopter. Unable to carry the 984 radar and retain general purpose capability they were fitted with pioneering digital data links to operate with the carriers. The programme was eventually cut back to eight, the second group of four carrying *Seaslug* Mk2 and a digital combat data system of their own. New general-purpose frigate programmes also began to replace the ships designed for specialized convoy escort roles. These new ships, First Rate Leanders and Second Rate Tribals, were designed primarily as task force ASW escorts and 'Gulf gunboats' respectively. They carried small *Wasp* helicopters, primarily for ASW weapons delivery.

It was in the Middle East, where a Rear Admiral at Bahrain now commanded the Gulf deployment, that the first major East of Suez emergency took place. The operation in support of Kuwait in 1961 allowed *Bulwark*, the first of the commando ships, to demonstrate its effectiveness. In the aftermath, the Chiefs of Staff carried out a study which placed even more emphasis on the importance of intervention to ensure the correct kind of successor regime as decolonization gathered strength. The 1962 Naval Estimates revealed almost total commitment to expeditionary operations by identifying no other role of the Navy than the following:

> In peacetime the ships of the RN are stationed all over the world. But when danger threatens they can be quickly assembled to take their place with the Army and Royal Air Force in combined operations to meet the threat. Every ship has her part to play. The commando ships and assault ships put ashore the spearhead of the landing forces with their guns, tanks and vehicles. The aircraft carriers provide reconnaissance and tactical strike ahead of the landing; air defence for the seaborne force; and close support for the troops ashore – especially when this cannot be done, either adequately or at all, by land based aircraft. Cruisers and escorts reinforce the air and anti-submarine cover, direct our aircraft and give warning of the enemy's and use their guns for bombardment if required. Submarines provide additional protection against hostile submarines and carry out reconnaissance and minelaying. The minesweepers clear a way to the land.[17]

Amphibious operations seemed much more natural to most naval officers than the new role that was somewhat reluctantly accepted at the end of 1962 – that of the deployment of a submarine strategic nuclear deterrent. Submarines of a more conventional type had always been important in the post-war navy. The need to operate close to a hostile Soviet coast and to exercise British surface ASW escorts led the operational force to rise to a peak of over forty boats in the mid-1950s. Despite experiments with hydrogen peroxide, the more conventional fast battery electric drive was chosen for post-war conversions and new construction, although the latter was delayed and the first completely new Porpoise class fast battery drive boat did not enter service until 1961.

By that time, a nuclear-powered submarine programme was well under way, helped by the strong advocacy of Mountbatten, to whose enthusiasm for technical novelty and innovation the idea appealed. Considerations of prestige were also important. The nuclear submarine was being referred to as the new capital ship. A navy with Britain's

---

[17] Cm 1629, paragraph 2.

traditions could not forego sharing in this vital new dimension of naval warfare. The help of the USN was enlisted and, after a complete propulsion system had been purchased from the United States, HMS *Dreadnought*, Britain's first SSN, was at sea on her trials before the end of 1962. An all-British nuclear submarine, *Valiant*, was already under construction. The SSNs could even be fitted into the East of Suez context, given their long range and the delivery of large surface ships to potential enemies such as the Indonesians.

Expertise in nuclear submarines also meant that the option was there to put the British deterrent to sea, where it was less vulnerable to pre-emption. This appealed to Mountbatten, but both he and the Service as a whole had reservations. The RN had been at the forefront of the 'nuclear sufficiency' debate, arguing that, in a world of mutual nuclear deterrence, scarce defence resources should be spent on usable limited war forces rather than over-provision on thermonuclear striking force.[18] An SSBN programme would be a diversion. It would also tread firmly on the toes of the RAF, a Service the Navy wished to see as fully committed as possible to the role of strategic bombing – for the time being at least. If the junior Service lost strategic bombing, its competition with the Navy for the role of providing Britain's limited war air resources would become more intense.

With a carrier replacement an increasingly pressing issue, the Navy wished to avoid such a conflict, especially as naval airmen were becoming a dominant faction in the East of Suez Navy. When, to the relief of the Admiralty, the expensive *Blue Streak* Intermediate-Range Ballistic Missile (IRBM) was cancelled the navy was happy to see it replaced by the *Skybolt* air-launched ballistic missile. The missile *Polaris*, it was thought, would be a logical follow-on to the latter as a limited national contribution to the overall Western deterrent. Some preliminary planning took place but the Naval Staff wished the carrier question to be settled first. Then, in late 1962, Washington cancelled *Skybolt* and *Polaris* was the only realistic alternative. Its procurement was agreed at a meeting between Prime Minister Macmillan and President Kennedy in Nassau in December.

This meant that by the early 1960s, the RAF and the RN had become competitive suppliers of tactical air power. The RAF had no other role whereas the RN had others upon which it could fall back. The stakes

---

[18] See, Richard Moore, *The Royal Navy and Nuclear Weapons* (Abingdon, OX: Routledge, 2001).

for the former service in this struggle were thus rather higher. The RAF's response was to formulate the 'Island Stance', a scheme of bases to provide a land-based alternative, supplemented perhaps by smaller carrier-type vessels to act as forward bases for the P1154 short take off/vertical landing (STOVL) fighter. By 1964, this competition was to take a new course as the context of British defence policy-making changed fundamentally. First, in April 1964 the three separate service ministries were combined into a unified Ministry of Defence and the Board of Admiralty ceased to exist. The latter was replaced within the new ministry by an 'Admiralty Board' subject to the overall Defence Council.[19] Six months later, Labour resumed office after thirteen years in opposition, committed to a self-consciously reformist line. The new government inherited a defence budget that was steadily rising but which, nevertheless, could not fully cover commitments.

Sharing the Burden of National Defence

The Wilson administration and its strong-minded Defence Minister, Denis Healey, were determined to produce a healthier relationship of inputs to outputs in defence policy, while reducing the burden of defence on an unhealthy economy. The government as a whole was not yet willing to withdraw from East of Suez but it wished to fulfil its peacekeeping and limited war responsibilities in the most cost-effective manner possible. The RN and RAF put their respective cases, carrier versus the Island Stance. The new bureaucratic structure picked out the weaknesses in the navy's forensic skills. Healey was persuaded by the apparent strength of the RAF's case. The Island Stance was adopted and the navy's carrier project known as CVA01 (for which the name Queen Elizabeth had been chosen but not published) cancelled. Although the carrier *Ark Royal* was to be given a major refit to sustain a carrier force as long as possible, the RAF would provide the air support East of Suez when the carriers disappeared.

The new First Sea Lord, Sir Varyl Begg, a gunner and therefore the archetypal 'surface' sailor, presided over a major reassessment of the navy's future role, size and shape. In April 1966, a Future Fleet Working Party was set up. It explored a number of studies, most notably an escort cruiser. Such a vessel, designed to carry anti-submarine

---

[19] Although a very different constitutional animal, the new Board's papers are filed in the National Archives with the Board of Admiralty's at ADM 167.

helicopters and surface to air missiles, had been considered since 1960, either as a complement to carriers East of Suez, relieving them of the necessity to carry ASW aircraft, or as the largest ship required to fight submarines in the Atlantic.[20] The Type 82 was envisaged as a possible solution, though it was eventually acknowledged that, as an escort, it was too large and expensive and the programme was cut back to one ship, HMS *Bristol*. The Working Party explored a number of destroyer, frigate and patrol vessel studies of various sizes from 1,200 to 4,500 tons.[21] Its report was briefed to Healey and a Board sub-committee appointed to consider its findings. By mid-1967 three new classes of ship were being planned, the cruiser, a smaller *Sea Dart* destroyer and a new *Leander* successor frigate.

The Working Party had pointed out that the government's planned budgetary allocations were insufficient to meet commitments both East and West of Suez. The Government agreed. In 1967, it was decided first to halve expenditure East of Suez and then to withdraw by the mid-1970s, although a 'special capability', including perhaps naval and amphibious forces, would be retained for use in the area. This compromise lasted only a few months. The economic situation worsened rapidly in the second half of 1967 and, in the aftermath of devaluation in January 1968, it was decided to withdraw from East of Suez by 1971, and withdraw all the carriers in that year.

Given the salience of East of Suez to the RN's rationale for survival, these events posed a serious challenge, but a new First Sea Lord, Sir Michael Le Fanu, and Navy Minister, David Owen, provided dynamic and innovative leadership. A new role was soon asserted for amphibious forces on the flanks of NATO and the commando force suffered only marginal reductions. Anti-submarine warfare was given a new lease of life, along with the British contribution to the NATO Striking Fleet. In 1968 an 18,700-ton, nine helicopter 'through deck design' was chosen for the ASW cruiser.

Naval forces were to form a major part of the 'general capability' that would be retained for areas outside NATO. Nevertheless, the major role of the RN would be found closer to home in the context of the Alliance's new strategy of Flexible Response. As a contemporary statement put it:

> Our withdrawal from overseas will enable us to increase the number of ships at immediate readiness for NATO's shield forces, and so enable

---

[20] Brown and Moore, *Rebuilding the Royal Navy*, 61-8.
[21] Ibid., 90-3.

us to continue to play a leading part among the European navies in the NATO maritime alliance ... The growth of Soviet maritime strength ... has underlined the importance of the shield forces especially in relation to the flanks of Europe, Scandinavia and the Mediterranean, where the increase in the Soviet naval presence has been most evident. Such shield forces were not, it was argued, intended to fight a full-scale battle of the Atlantic, but they would 'identify aggression when it occurred' and prevent it developing into a more serious conflict. [22]

Despite the election of a new Conservative Government in 1970, the withdrawal from East of Suez duly went ahead in 1971 and the following year the deployment of operational major surface combatants and amphibious ships looked very different from what it had been eight years before. In 1964, there had been twenty-eight assets in European waters, thirty-five East of Suez, two in the South Atlantic at Simonstown and two in the West Indies. In 1972, there were fifty-one in European waters, six East of Suez and three in the West Indies (the South Atlantic Station had been abolished in 1967). There had been only a 10% reduction in total strength.[23]

During the 1970s a new fleet began to take shape to replace the old. The gun-armed destroyers finally disappeared from service and the Type 42 *Sea Dart* destroyers began to be delivered to replace them. Their all gas turbine propulsion had been validated by trials in the converted Type 14 *Exmouth*. Two new frigate programmes were eventually adopted. One was the relatively cheap but rakish Type 21, the *Amazon* class, largely designed by private industry due to overloading of the Ministry of Defence's own designers. The other was the larger, more expensive Type 22 ASW ship, the definitive replacement for the *Leander*. The Type 23 was delayed until 1975 by problems with its advanced *Sea Wolf* short range anti-missile missile and two extra Type 21s were built instead, bringing total numbers to eight, delivery of which took place in 1974-78. Delivery of *Leanders* ended in 1973, but the class began to be modernized with *Ikara* ASW or *Exocet* surface-to-surface missiles. The latter were procured to make up for the limited anti-surface strike potential of the post carrier navy; they were also fitted to the four later Counties and the new frigates. The SSN programme also went steadily ahead after the hiatus caused by the construction of the Polaris force. A new 'S' class boat of improved

[22] House of Commons debate, 11 March 1968, columns 1008-9.
[23] Deployment figures from *Navy International*, March 1964 and February 1972.

design was developed to supplement the five boats of the Valiant class. The first was laid down in 1969 and commissioned in 1973.

It was large surface ships, however, which seemed to provide a more obvious index of naval capability. There was much rejoicing, therefore, when the first of the through-deck command cruisers, HMS *Invincible*, was ordered in 1973. At that time it was still not entirely clear whether this class would operate the P1127 production derivative, the *Harrier* STOVL fighter, the decision being delayed by the government's political and economic problems, which culminated in its premature fall. The returning Labour administration, however, finally ordered a special *Sea Harrier* for *Invincible* and two extra ships of the class that were also laid down. In the meantime, the commando carriers were given an ASW role to fill the gap between the withdrawal of *Ark Royal* and the delivery of the Invincibles. Despite much rhetoric of Defence Review and cuts in expenditure, expenditure on the Navy increased under Wilson and his traditionally pro-Navy successor, James Callaghan. In 1978-79, as much was being spent on Naval General Purpose Forces in real terms as in 1966-67.

From the 1975 Defence Review to the Aftermath of the War in the Falklands

The Labour Defence Review of 1975 officially liquidated naval commitments outside the Eastern Atlantic and Channel areas. The last British warship finally sailed from Malta in 1979, although Hong Kong and the West Indies remained significant commitments outside European waters, as did the Falkland Islands with its ice patrol vessel, the retention of which was insisted on by the Foreign Office. Group deployments also continued to go on world cruises to demonstrate the global dimension of the residual general capability.

The Thatcher administration elected in 1979 had enormous problems managing its defence budget so a new more effective Defence Minister, John Nott, was appointed to sort out the situation. Nott recognized that the planned programmes of the services were too large to be fitted within any expected longer term defence budget. The Callaghan government had signed up to a three per cent increase in real terms. This had been endorsed by Thatcher but, as Nott put it in his memoirs: 'No one in the government when I joined the MOD had suggested that the 3 per cent annual growth target might last beyond 1983-4, yet public commitments to the equipment programme were being given on the assumption that 3 per cent volume growth would continue until 1989-90.' This would have 'bankrupted the exchequer'.[24]

Nott was determined to 'design a force structure to meet the main threat to the United Kingdom – and to make that force structure sufficiently flexible to meet the unexpected'.[25] The Navy briefed Nott on the existing convoy/striking fleet strategy and its demands but when the Secretary of State went to meet Supreme Allied Commander Atlantic he was informed of the latest thinking about 'a series of hectic, dispersed single sailings to avoid the Soviet submarine threat, rather than Second World War escorted convoys'.[26] Both within NATO and American circles, the idea was to move towards a less reactive policy of defence in depth through containment and to keep the initiative. Forward operations by SSNs and submarines would contain the Soviet fleet by distraction and maintain the ability of NATO reinforcement shipping to cross the Atlantic via the Azores lightly defended or even unescorted.

Unfortunately, despite the adoption of a new NATO CONMAROPS along these general lines there was not yet the unity of Alliance views that prevailed in the mid-1980s. All Nott obtained from his American visit was reason to doubt the advice he was getting from his own dark blue advisers. As he later put it, '(w)hat was needed was fresh and original thinking to meet the Soviet submarine challenge; I got none. I was forced to seek proposals elsewhere and impose my own priorities within the growing naval programme.'[27]

Nott used his central Defence Staff to prepare a joint paper on 'the effectiveness of different naval assets against the Soviet submarines'. Based on the work of the Defence Operations Research Establishment at West Byfleet, this study placed the carrier-based aircraft/Sea King combination in the fourth category of effectiveness, near the bottom of the list. Nott claims that if he had not been making the Soviet threat the first priority he would have endorsed the full *Invincible* class programme of three ships. As he put it, 'I did not see how we could afford three carriers in planning for a high intensity war against the Soviets,

---

[24] John Nott, *Here Today Gone Tomorrow: Memoirs of an Errant Politician* (London: Politico's Publishing, 2002), 108.
[25] Ibid., 211.
[26] Ibid.
[27] Ibid., 212.

nor how at that time we could afford to equip ourselves, in priority for out of area low intensity war – not least because the protection of the carriers required a flotilla of supporting frigates of which we had few in number anyhow.[28] Nott decided to maintain a carrier force of two ships.

It must be stressed, given common misconceptions about Nott's carrier policy, that his intention was to maintain a force of two ships to provide some 'out of area' capability. The government had emphasized this additional dimension of defence policy in its statements and Nott fully accepted that a substantial surface navy, as much as could be afforded after other more important commitments were met, was the key to maintaining a credible residual capacity beyond the Tropic of Cancer. The naval priority was to modernise the nuclear deterrent in the future form of *Trident* and to keep the Eastern Atlantic role by a revised posture emphasizing submarines, Nimrods and new austere towed array frigates. The decision was taken to put the costs of the strategic nuclear force into the overall 'naval' budget a factor that increased pressure on the rest of the Naval programme.[29] The key question was how much of the remaining surface fleet could be fitted into the planned budget?

Nott's Under-Secretary for the Navy, Keith Speed, had just done a study of the naval dockyards. The Speed Report argued that the current policy of giving ships substantial refits was not cost effective. The only reason to keep it was the political one of maintaining three major dockyards. Nott was willing to bite this particular political bullet. It was decided that Chatham dockyard would close and the fleet be cut back to a size that could be maintained by the remaining dockyard infrastructure that would also be reduced by a rundown at Portsmouth. This would mean a reduction in the NATO declaration of frigates and destroyers from fifty-nine to fifty of which eight would be in standby reserve. The Royal Fleet Auxiliary (RFA) fleet was to be reduced by two tankers and a store ship and the amphibious ships were to be withdrawn. The amphibious squadron would be reduced to the RFA manned Landing Ships Logistic.

In contrast to the planned reductions, Nott's supplementary White Paper 'The Way Forward' in a section that was lost in the protests of its supporters clearly stated that the RN had a 'particularly valuable role'

---

[28] Ibid., 229.

[29] Nott insisted on this both in his book and at the Joint Services Staff and Command College (JSCSC) Witness Seminar; it was contested at the latter event by his former Permanent Secretary, Sir Frank Cooper, on basis of denying the existence of a single 'navy' budget.

in 'efforts' outside the NATO area, for which British 'needs, outlooks and interests gave' her 'a special role and special duty'. It announced the intention 'to resume from 1982 onwards the practice of sending a substantial naval task group on long detachment for visits and exercises in the South Atlantic, Caribbean, Indian Ocean and further east'. The intention was nevertheless announced to decommission the two assault ships *Fearless* and *Intrepid* over the next two years but the decision to go to Trident D5 rather than C4 later allowed a rearrangement of the programme, releasing funds for their retention. Officials in Washington had also raised anxieties about the disappearance of the amphibious assets.

The Argentine invasion of the Falkland Islands may well have been prompted by a small but significant part of the 1981 Review, the withdrawal of the ice patrol ship *Endurance*. Two Task Forces, one of carriers, amphibious ships and ground forces, Task Force 317, and one of submarines Task Force 324, were sent to the South Atlantic and they successfully recovered the islands.

Despite the commonly held view that the Falklands War led to a major re-appraisal of Nott's policies, there is a good case to be made that, although the war did have some marginal effects on policy, the basic direction mapped out in 1981 was not greatly altered. To cover the four ships lost in the Falklands, two more Batch 2 Type 22 frigates were ordered while three older destroyers (including HMS *Bristol*) slated for early disposal would be retained. The definitive replacement of the four lost ships proved to be four enlarged Type 22s, fitted unlike their predecessors with a gun armament. The new Type 23 'simpler and cheaper type of anti-submarine frigate' became a 'general-purpose frigate' also with gun for shore bombardment and a helicopter. A third carrier was maintained.

The frigate/destroyer fleet was not to be cut to fifty ships (of which eight would be in reserve in the Standby Squadron). Instead, numbers would be retained at 'about 55' through to April 1984. In addition to the destroyers already announced for retention, three un-modernized *Leander* class frigates, three Leanders modified with *Ikara* ASW missiles and an old Type 12 frigate were reprieved. Improved point defences would be fitted to all the carriers, the amphibious ships, HMS *Bristol* and the Type 42 destroyers. Airborne early-warning helicopters were to be provided for each operational carrier.

The 1985 Defence Statement confirmed that Nott had been right and 1985-86 would indeed be the last year of the three percent increase.

From 1986 to 1987 when the costs of defending the Falklands were taken into account, the planned increases in defence expenditure were barely ahead of the inflation rate. This was not enough to maintain the strength of the RN in either personnel or ships. Personnel numbers were reduced to less than 70,000 in 1985. The fifty-five frigates and destroyers of 1983 were reduced to fifty-two front-line ships in 1984, the year in which it was announced that the total force would amount to Mr Nott's fifty units, albeit all in commission. The following year, front-line numbers had dropped to forty-six and in 1986 the words 'about 50' were being used to describe the force level. Nevertheless, frigate and destroyer numbers dropped further to forty-one in 1988, only marginally more than Nott's FF/DD flotilla. Claims that the Falklands somehow 'saved' the surface fleet thus need revision.

The Disappearance of the Soviet Threat and the End of the Cold War

Until the end of the Cold War, the main gain of the 1980s was the extra carrier, but the impact even of that was mitigated by the maintenance of only two air groups. The veteran *Hermes* was taken out of service in April 1984, over a year and a half before its replacement, the new carrier *Ark Royal* was commissioned. The third carrier would always be at considerable notice when in reserve or extended refit. A major debate also went on over the replacement of specialist amphibious shipping. A 'firm decision' to keep the amphibious assault role was eventually announced in 1986 but little was done in the short term to put this into effect.

The case for these ships was helped by the renewed emphasis on Northern Flank operations with the adoption of the US Forward Maritime Strategy that influenced the way CONMAROPS was interpreted. The result was an unprecedentedly coherent doctrine of 'forward operations'. Allied nuclear powered attack submarines would penetrate the 'defended bastions' where lurked the Soviet ballistic missile firing boats, to tie down Soviet assets in their defence. The NATO carrier Striking Fleet with a new RN commanded 'Anti-Submarine Striking Force' of towed array surface assets led by a British ASW carrier would cover amphibious landings in Norway and so threaten the Soviet Northern Fleet that its main striking assets, both submarines and land based bombers, would be drawn out and destroyed. During the mid-1980s this concept was tested and developed in NATO exercises.[30] The NATO Striking Fleet also began to exploit the Norwegian fiords as 'bastions' from which to take on the Soviet Northern Fleet.

The RN remained balanced between important roles in the Atlantic and Channel and, if necessary, for use further afield. As Admiral Fieldhouse, now Chief of the Defence Staff put it in 1986, '(W)e will not be able to devote substantially greater resources directly to our out-of-area capability than we do today. We must therefore ensure that what we do is best tailored to our needs, and that we engineer flexibility and mobility into our NATO forces wherever possible so that double earmarking makes sense.'

The diminished threat from Gorbachev's Soviet Union was also making the overwhelming preoccupation with the Soviet and Warsaw Pact threats look a little misplaced. Out-of-area exercises were beginning to take more prominence, notably 'Saif Sareea' in and off Oman in 1986, supported by the aircraft carrier *Illustrious* and the Global 86 Task Group, the first RN Task Group to circumnavigate the globe for a decade. In 1987, a major joint service exercise held in and off Scotland, named 'Purple Warrior', was designed to test operations against a fictitious island group 1,500 miles from the UK. This exercise involved some 20,000 personnel and thirty-seven ships and was among the largest amphibious exercises carried out by Britain since the end of the Second World War.

The first stirrings of a more maritime strategic emphasis for the United Kingdom were thus talking place even before the Berlin Wall came down in 1989 and the Soviet Union collapsed at the end of 1991. Although the main force of the Major Government's review fell on the continental commitment of ground and air forces, the RN was forced to take its share of the post-Cold War cuts. It shed all ten of its remaining relatively recently modernized *Oberon* class conventional submarines only leaving the four new conventional *Upholders* to accompany the dozen 'S' and 'T' class SSNs. The frigate and destroyer force was reduced officially to 'about 40'.[31]

Naval forces played a significant role in the Gulf War of 1991 and its aftermath but despite the changed strategic environment there was a struggle over the 1993 Long Term Costings, a type of mini-Nott review

---

[30] Se Eric Grove and Graham Thompson, *Battle For The Fiords: NATO's Forward Maritime Strategy in Action* (Annapolis, MD: Naval Institute Press, 1991).

[31] Cm 1981, Statement on the Defence Estimates 1992, 9.

with continued investment on European mainland (in the new form of NATO's Allied Rapid Reaction Corps) set against deployable maritime capability. The planned new amphibious helicopter carrier, an addition to amphibious capability to allow a two company helicopter lift and for which tenders had been submitted in 1989 but been allowed to lapse, only escaped cancellation by the skin of its teeth. The RN had, however, to sacrifice all its conventional submarines and accept 'about 35' frigates and destroyers. The new amphibious ship, HMS *Ocean*, was laid down in 1994.

Financial pressure continued to increase as the Treasury revised downwards its annual planned defence provision, to little over £22 million for the mid-1990s. As the First Sea Lord Admiral Sir Benjamin Bathurst put it in retrospect:

> The choice which the three services faced was quite clear: either front line could be replaced by finding savings elsewhere, or the challenge could be ducked, with the result that fighting capabilities would be diminished. The decision to place the front line first was entirely the right one, and it was the course of action which accorded exactly with long held Navy Board practices.[32]

Nevertheless, the Defence Costs Study established at the beginning of December 1993 to meet the budgetary shortfall found it possible to make further significant savings. As 'sweeteners', plans to enhance the front line were announced simultaneously with the Study, notably an examination of the arming of SSNs with cruise missiles and the completion of the *Sandown* class single role mine-hunter programme. Five of these cheaper MCM vessels had been completed before the Ministry had cancelled tenders in 1991. Now all the last seven projected assets were to be ordered to bring the total number of MCM vessels up to the planned total of twenty-five.

The strength of the RN and Royal Marines was 45,233 at the beginning of 1997.[33] In these circumstances it was more important than ever to replace older ships by more manpower intensive assets. By the end of 1996, there were eleven Type 23 'Duke' frigates in service, each requiring only 174 officers and men, compared to about 250 or more in a Type 22 or Type 42. The six remaining Type 21 warships were sold to Pakistan in 1993-94 and the first batch of Type 22 units soon followed

---

[32] Broadsheet, RN Annual Publication 1994-95, 2.
[33] Figures from contemporary *Janes Fighting Ships*.

them out of service, being sold to Brazil in November 1994. The last unit of the *Leander* class, the modernized HMS *Scylla*, was decommissioned in 1993. Another important development in the mid-1990s was the replacement of the *Polaris* submarines by the new *Trident* vessels. The impressive 16,000 ton *HMS Vanguard* began her first operational patrol with sixteen Trident D-5 missiles at the end of 1994. The arrival of *Victorious* in 1995 allowed the withdrawal of the last *Resolution* class Polaris SSBN, HMS *Repulse*, the following year (the class had begun paying off in 1992).

The break up of Yugoslavia allowed the RN to demonstrate its key roles in the new world disorder. As peacekeeping forces were deployed ashore a carrier with its Sea Harriers was kept on station from the beginning of 1993 to provide a national contingency capability to support British contribution. The Government regarded its presence as a condition for the presence of a British component in UNPROFOR.[34] The lesson of possessing a degree of power projection was not lost on the new government when, in 1997, the era of Conservative government finally ended with the election of a 'New Labour' government under Tony Blair. George Robertson was Secretary of State for Defence and he immediately embarked on the *Strategic Defence Review* (SDR). The result in 1998 was a triumph for the RN. Britain's strategic horizons were widened to include not just Europe but the surrounding areas of the Mediterranean and the Gulf (back East of Suez once again). The strategic implications of this for the UK's maritime posture were made clear in George Robertson's Introduction: 'In the post-Cold War World, we must be prepared to go to the crisis, rather than have it come to us. So we plan to build two new larger aircraft carriers to project power more flexibly around the world.'[35]

Recent experience in the Gulf was also quoted. There, as a result of Saddam Hussein's non-compliance with the post-1991 armistice agreement, HMS *Invincible* was deployed in early 1998. Her squadron of *Sea Harrier FA2* fighters (the new electronically and weapons enhanced variants in service since the mid-1990s) were reinforced by a group of RAF *Harrier GR7*. She was then sent to operate with the US carrier

---

[34] Personal experience. See the author's book *Ark Royal* published by the ship in 2002.
[35] UK Ministry of Defence (MoD), *The Strategic Defence Review* (London: MoD, 1998), Cm 3999, 2, http://www.mod.uk/NR/rdonlyres/65F3D7AC-4340-4119-93A2-20825848E50E/0/sdr1998_complete.pdf (Accessed on 23 November 2010).

battle groups threatening Iraq. Carriers were particularly important in this situation as host nation support for Anglo-American operations was difficult to obtain. Operation Bolton provided a timely reminder of the utility of carrier basing in the power projection role just as the SDR was being drawn up. The joint air group presaged the announcement in the SDR of Joint Force 2000, a new grouping of Sea Harriers and Harriers as a joint force. This inter-service agreement was a necessary precursor of the whole new carrier programme.

The Navy did best out of the revised SDR force posture. With carrier replacement promised, it was willing to reduce the submarine force to ten, the destroyers and frigates to thirty-two (sixteen type 23s, five Type 22s and eleven Type 42s) and the mine warfare vessels by three to twenty-two. More crucially, the SDR confirmed the importance of the amphibious ships. The month before the SDR appeared in June 1998 the first of the new landing platform dock (LPD) ships, HMS *Albion* was laid down at Barrow; her sister *Bulwark* followed in 2000. With the coming into full service of *Ocean* in 1999 the opportunity was taken to put *Intrepid* for disposal leaving *Fearless* to soldier on until the two new ships appeared, which they did in 2003 and 2005. Taken together, this amphibious squadron, including the landing ships operated by the RFA, was as powerful as any deployed post-1945.

The SDR announced that all the SSNs would be fitted with Tomahawk land attack cruise missiles. The boat HMS *Splendid* fired its first missiles in October 1998 and *Spartan* was fitted in 1999 with two 'T' class boats being equipped in 2000. The 1997 contract to build three new larger SSNs with more weapons capacity was continued and first steel was cut in 1999. The RN now clearly saw power projection as its basic role with the major platforms, carriers, amphibious ships and SSNs enabled by the frigate/destroyer force and the MCMs.

Conclusions

As the United Kingdom entered the twenty-first century, defence policy was to undergo some changes, but none that would question the balance of the fleet as it had emerged from the end of the Cold War. In July 2002, a 'New Chapter' sought to address the impact of 9/11 on British defence policy, and subsequent operations in Afghanistan and Iraq confirmed, if anything, that the expeditionary focus would become even more pronounced. Engagement might well be required 'further afield more often than perhaps we had recently assumed'.[36] This was confirmed in the 2003 White Paper which specifically added 'areas where we have strong historical ties and responsibilities', Sub-Saharan Africa (a successful maritime intervention had been carried out in Sierra Leone in 2000) and South Asia.[37] The UK had a global strategy once more.

Unfortunately, the Government was still under pressure to limit defence spending and under the pretext that network enabled capability allowed fewer platforms to be deployed in 2004 three destroyers were cut, three Type 23 frigates and six mine countermeasures vessels.[38] The SSN force was cut to eight boats allowing about five to be kept running at all times. The new *Astute* submarine programme was troubled by delays but by the end of the decade four boats were at varying stages of construction to replace life expired SSNs, with *Astute* herself about to enter service.

The main item, however was the carrier programme. This has gone ahead with contracts signed that will be expensive to cancel. Construction began in 2009. The completion of HMS *Queen Elizabeth* and HMS *Prince of Wales*, at 65,000 tons was to provide the RN with the largest ships ever built. These platforms were to confirm the RN as a world class exponent of maritime power projection and enable the United Kingdom to retain a fleet second only to that of the USA in reach and overall capability. Given the manifold financial problems faced since the end of the Second World War this was no mean achievement. The history of British post-war defence policy is one of constant struggle between regional and global interests, with sea power representing a key enabler of foreign and security policy. In key tests, in the Falklands, in the Gulf, in the Balkans, as well as in places like Sierra Leone, the flexibility of naval platforms enabled the country to project power and achieve national political objectives. Whilst constantly shrinking in numbers, the RN endeavoured to keep a degree of balance in its force to enable the country to remain a core player on the international scene.

---

[36] UK Ministry of Defence, *The Strategic Defence Review: A New Chapter* (London: MoD, 2002), Cm 5566-1, 12, http://www.mod.uk/NR/rdonlyres/79542E9C-1104-4AFA-9A4D-8520F35C5C93/0/sdr_a_new_chapter_cm5566_vol1.pdf (Accessed on 23 November 2010).

[37] UK Ministry of Defence, *Delivering Security in a Changing World: Defence White Paper* (London: MoD, 2003), CM 6041-1, 5, http://www.mod.uk/NR/rdonlyres/051AF365-0A97-4550-99C0-4D87D7C95DED/0/cm6041I_whitepaper2003.pdf (Accessed on 23 November 2010).

[38] On the topic, cf. UK Ministry of Defence, *Delivering Security in a Changing World: Future Capabilities* (London: MoD, 2004), Cm 6269, http://merln.ndu.edu/whitepapers/UnitedKingdom-2004.pdf (Accessed on 22 November 2010).

CHAPTER SIX

PUNCHING BELOW ITS WEIGHT: JAPAN'S POST-COLD WAR EXPEDITIONARY MISSIONS

Chiyuki Aoi

Most of the military operations that Western nations conduct today are expeditionary in nature and, most expeditionary operations are stability missions. Cases in point are the two major expeditionary counterinsurgency operations proceeding (as of early 2011) in Afghanistan and Iraq. Whether or not one agrees with the proposition that such expeditionary small wars – not the major wars of the type fought on European soil in the twentieth century – are now the central security concern, there can be no doubt that these two particular operations will continue to absorb resources for some time to come. Indeed, the post-Cold War security environment has dramatically altered the strategic value of what used to be called 'low-intensity operations' – a shift that reflects practical realities in a time when instability and related contingencies directly impinge on a nation's security.

The purpose of this chapter is to analyse Japan's response to this shift and to evaluate the expeditionary capabilities of the JSDF, especially in relation to stability operations. It aims also to discuss the challenges and obstacles that Japan faces in expanding these capabilities. The chapter addresses the question of how and to what extent Japan has developed expeditionary capabilities in the post-Cold War era. The country conducted no expeditionary missions at all during the Cold War (the only foreign mission was the mine-sweeping mission that the then-occupied Japan conducted during the Korean War) – such actions were virtually taboo. By contrast, in the post-Cold War period, the number of missions by the JSDF has increased dramatically, as Japan started to engage in UN PKOs and Special-Measures Law-related activities, and later anti-piracy operations, in addition to international disaster relief operations.

One of the main arguments of the chapter is that for the large part of the post-Cold War period Japan met the new demands of expeditionary missions largely by utilizing existing equipment and force structure.

It did so to balance these new missions with the need to accommodate other interests, including the maintenance of the regional balance of power and attention to domestic disaster relief. A number of institutional upgrade packages together with efforts to update and expand certain equipment, however, started to enhance Japan's expeditionary capabilities since the latter half of the 2000s.

A second goal of this chapter is to identify and explain factors that continue to constrain Japan's expeditionary capabilities. This is despite Japan having the potential in terms of military might to develop major expeditionary capabilities. Indeed, in terms of hardware, Japan has achieved a middleweight expeditionary capability since the end of the Cold War. On the other hand, it still lacks the 'software' capabilities appropriate for expeditionary missions, especially in terms of operational framework and doctrine, under-girded by relevant experience and advanced education and training in relevant areas. Japan's expeditionary 'international peace cooperation' missions have so far been confined to a certain range by self-defined and self-imposed conditions and limitations that have marginalized Japan's contribution to international missions for enhancing stability.

Such limitations stem in part from legal prohibitions on JSDF use of force abroad. They are also related to Japan's own ambivalence in determining what national interests are involved in expeditionary-stability endeavours. It is, as a result, partly a problem of how to define the role/place of expedition and stabilization within the national security strategy. In this respect, Japan's expeditionary capabilities reflect the country's long-standing security dilemma. On the one hand, Japan's conservative-realist political orientations have led politicians to seek to limit the nation's external security role based upon parsimonious considerations in terms of human and material costs. On the other hand, Japan pursued an 'ideological multilateralism', an *ad hoc* form of multilateralism aimed at securing a permanent membership on the UN Security Council. With regard to the legal, operational, and doctrinal constraints, the currently persuasive theses emphasizing how far Japan has come during the last two decades towards becoming a more active international player or 'normal' nation (that is, one that participates in international military-security actions and/or engagements like the other industrialized democracies) are probably misleading. Japan has changed little in terms of overcoming the post-war legal and political prohibitions on external engagement.

The following sections focus on the emergence and development of Japan's expeditionary missions and capabilities through the Cold War and the post-Cold War eras. The discussion then turns to analysis of operational limitations, with particular attention to conditions for JSDF deployment abroad and constraints emanating from the JSDF mandate and authority. Afterwards, the chapter engages in an examination of operational doctrine applicable to JSDF expeditionary-stability missions, deduced from past practices, which sheds light on the disparity between Japanese 'doctrine' (or operational principles) and international standards today. A few final considerations will examine the widening gap between Japan's avowed political ambition of playing a larger security role and its actual defence capabilities – especially its capabilities in the software dimension.

Defining Expeditionary and Stability Missions

The UK Ministry of Defence defines expeditionary operations as 'military operations, which can be initiated at short notice, consisting of forward deployed or rapidly deployable self-sustaining forces tailored to achieve a clearly stated objective in a foreign country'.[1] Although this definition centres on the nature of force employed, as Geoffrey Till observed in his consideration of sea power, expeditionary missions are 'usually highly politicized', involving more than naval coercion and other simple exercises of military power.[2] Expeditionary missions are politicized in that they serve specific aims set by the foreign policy of the intervening nation. By definition, they are conducted away from the homeland and are thus often campaigns of choice, typically under the (often false) assumption of limited costs. As such, expeditionary missions are often joint ones, with naval, ground and air forces playing crucial roles depending on the type of deployment.[3] Hence, any assessment of 'expeditionary' capabilities must also entail looking at the capabilities of the three armed services as well as at joint capabilities.

---

[1] The UK Ministry of Defence, *UK Doctrine for Joint and Multinational Operations (UKOPSDOC) JWP 0-10* (Swindon: The Development, Concepts and Doctrine Centre, September 1999), cited in Geoffrey Till, *Sea Power: A Guide for the Twenty-First Century* (London: Frank Cass, 2006), 235.

[2] Till, *Sea Power*, 237.

[3] Ibid.

It should be noted that the concept of expedition is distinct from 'power projection' in the case of Japan.[4] In most circumstances, there may be some overlap between the two, but in the case of Japan, the distinction is important. Japan's Constitution and other factors that limit its armaments make 'power projection' unlikely, no matter how sophisticated its expeditionary capabilities become as it strives to maintain order at sea, and less frequently, to secure international peace and security. It is important to recognize that such functions assigned to Japan's forces are without major power-projection potential.

Conceptualized as such, expeditionary missions naturally overlap with and are highly akin to stability operations, which are the main contingencies of the present era for armed forces, especially for industrially developed nations. Stability operations are currently defined by the US Army as 'various military missions, tasks, and activities conducted outside the United States in coordination with other instruments of national power to maintain or re-establish a safe and secure environment, provide essential governmental services, emergency infrastructure reconstruction, and humanitarian relief (JP 3-0)'.[5] Stability operations involve efforts related to reviving failed or fragile states. These tasks have a wide range: creating or strengthening state apparatuses and administration; addressing conditions of general instability, with or without a clearly identifiable 'enemy'; dealing with fundamental development challenges such as poverty, illiteracy, and poor health, sanitation and other infrastructure; capacity-building for sustainable economic growth and fiscal policy and control; and developing media and other infrastructure for a functioning society.

It is beyond the scope of this discussion to elaborate why and how stability missions came to occupy a central place in the national security

---

[4] I would like to thank Dr Alessio Patalano for impressing me with this point. See the paper Alessio Patalano, 'A New Sun on the Horizon? The Dawn of Japan's Expeditionary Capabilities', *Osservatorio dell'ISMM*, 2008. For a discussion of Japan's 'power projection', see Christopher W. Hughes, *Japan's Remilitarisation* (Adelphi Book, Abingdon, OX: Routledge for IISS, 2009).

[5] Quoted in US Department of the Army, *Stability Operations FM 3-07* (Washington, DC: Headquarters, Deprtment of the Army, 2008), vi–vii. See also the newly published UK Ministry of Defence, Security and Stabilization, *UK doctrine on Security and Stabilization, JDP 3-40* (Swindon: The Development, Concepts and Doctrine Centre, November 2009), which defines stabilization as 'the process that supports states which are entering, enduring or emerging from conflict in order to: prevent or reduce violence; protect the population and key infrastructure; promote political processes and governance structures which lead to a political settlement that institutionalizes non-violent contests for power; and prepares for sustainable social and economic development'.

policies of developed nations;[6] suffice it to say that stabilization is the primary purpose for conducting expeditionary missions today. Japan cannot remain unaffected by this general global shift that has brought the stability type of mission into the mainstream of strategic thinking.

The Emergence and Development of Japan's Expeditionary Capabilities

The idea of Japan conducting expeditionary missions was unthinkable during the Cold War. The 1947 Peace Constitution and its subsequent interpretations have made the country's participation in both expeditionary and stability missions extremely contentious. The terms of the Constitution render problematic the use of force by Japanese personnel for purposes beyond national defence based upon the right of individual self defence, either as an exercise of the right of collective self-defence or as part of multinational forces.[7] The dispatch of the JSDF abroad has been a controversial issue since 1954 when, after the JSDF Law was passed, the House of Councillors adopted a resolution (nonbinding) prohibiting the dispatch of the JSDF abroad.[8] The government subsequently interpreted this resolution to mean prohibition of overseas operations *for the purpose of the use of force*. That limitation on the role and mission of Japan's armed forces formed an integral part of the post-war national 'grand strategy' by which Japan was to focus on rapid economic recovery while the US-Japan security treaty provided for its security.[9]

Thus, expeditionary capabilities were not much of a concern during the embryonic stage of the JSDF and through much of the Cold War. The JSDF in its formative years of the 1950s and 1960s focused on domestic procurement of armaments, replacing the often old and obsolete armaments donated by or loaned from the US and increasing its firepower and mobility to a level capable of repelling direct attacks

---

[6] For extensive treatment of the issue, see Chiyuki Aoi, *Legitimacy and the Use of Armed Force: Stability Missions in the Post-Cold War Era* (Abingdon, OX: Routledge, forthcoming).

[7] On this topic, see General Yamaguchi's chapter in this book.

[8] Akiho Shibata and Yoshihide Soeya, 'Legal Framework', in L. William Heinrich Jr., Akiho Shibata, and Yoshihide Soeya (eds), *United Nations Peace-Keeping Operations: Guide to Japanese Policies* (Tokyo: United Nations University Press, 1999), 51.

[9] On the 'grand-strategic' status of this mercantilist policy pursued by Prime Minister Yoshida Shigeru, see Richard J. Samuels, *Securing Japan: Tokyo's Grand Strategy and the Future of East Asia* (Ithaca and London: Cornell University Press, 2007), 39–40.

on the archipelago. Responding in particular to the perceived increase of the Soviet threat in the Far East through the 1970s and in the 1980s, the JSDF continued to focus on national territorial defence, enhancing gradually both the amount and the technological aspects of its armaments, such as deployment of missile systems and deployment of increasingly sophisticated data and surveillance systems. As General Yamaguchi's chapter highlighted, no consideration was given to the possibility of foreign missions until the 1990s outside the context of sea lanes defence.

Then, for expeditionary missions, the Cold War stasis gave way to post-Cold War activism. Following the Gulf War in 1990-91, a mine-sweeping mission was conducted by the JMSDF in the Persian Gulf. This paved the way for Japan's new era of expeditionary missions. The JSDF participation in UN PKOs began in the aftermath of the international political humiliation Japan suffered when it failed to collaborate with the multinational forces in the Gulf War. UN PKOs constituted the core area in which Japan considered the use of expeditionary missions. Under the International Peace Cooperation (IPC) law passed by the Diet in 1992[10] Japan sent a 600-member engineer unit and eight military observers to the United Nations Transitional Authority in Cambodia (UNTAC) the same year. Between September 1992 and September 1993, two engineer battalions (totalling 1,200 personnel) and sixteen observers provided logistical support. Their tasks included aspects of logistics, like the repair of roads and bridges, the procurement of water and fuel for UNTAC units, of food and medical provisions, and sleeping and working facilities for UNTAC personnel. The JSDF personnel was tasked also with the securing of election-related goods, of providing transportation, along with monitoring of the cease-fire and disarmament. Provided the scale of the operation, air and naval assets were mobilized for the transport of personnel and equipment.

In Mozambique, the JSDF provided five persons to the UN Operation in Mozambique (ONUMOZ) headquarters and a forty-eight-member Transportation Control unit. In operations between May 1993 and January 1995, a total of ten JSDF personnel served in rotation at the ONUMOZ headquarters, and a total of 144 members served in the Transportation Control Unit. The JASDF also provided transport (of goods) to sustain the JGSDF (ground) personnel.

---

[10] The law has been amended twice, 1998 and 2001.

The Golan Heights mission is the one that JGSDF officers refer to as the 'PKO school', because the nature of the operation there involved typical peacekeeping duties, and the cumulative number of JGSDF personnel deployed was large. Today, it provides two officers to the UN Disengagement Observer Force (UNDOF) headquarters, as well as a forty-three-member transportation unit. The more limited nature of the mission entailed that the JASDF again provided transport of goods to the GSDF personnel. Since the start of the mission in February 1996, a total of 1,204 JGSDF personnel have served in the UNDOF Transportation Unit and thirty-two at UNDOF Headquarters.

Rather different, in terms of scale, was the JSDF operation in East Timor, the largest to date of Japan's participation in UN peacekeeping operations. Indeed, the mission had a significant role in the development of Japanese peace-building diplomacy. For the mission that started in February 2002, the JSDF dispatched ten officers to serve in the headquarters of the UN Transitional Administration in East Timor (UNTAET) as well as a 680-member engineering unit to serve as part of peacekeeping force there. Between February 2002 and June 2004, a total of seventeen officers served in UNTAET and the UN Mission of Support in East Timor (UNMISET) headquarters and a total of 2,287 persons served in rotation in the engineering unit. Both JMSDF and JASDF provided transport (of personnel and goods), with naval assets being indispensable for the transportation of heavy equipment. As part of the UN mission, JGSDF engineers provided logistical support and carried out repair of main supply routes (MSR), maintenance of water supply facilities, and repair and maintenance of UN facilities. Activities in the area of civil-military affairs intended to improve the lives of East Timorese were also an important element of the JGSDF mission. Being the only engineering unit deployed in East Timor, the JGSDF was also in charge of developing the UN's country-wide Civil Military Affairs (CMA) activities. This engagement represented a crucial precursor to the JSDF civil assistance in Iraq, although the methodologies of assistance differed significantly in the two locations.

Following the operations in East Timor, Japan assigned six unarmed JGSDF officers to serve as military observers at the UN Mission in Nepal (UNMIN) in March 2008 (still ongoing). It also posted two JGSDF officers to serve in the headquarters of the UN Mission in Sudan (UNMIS) in October 2009 (still ongoing).

In terms of unit-participation in UN PKOs, the JSDF involvement in the aftermath of the 2010 Haiti earthquake filled a long hiatus after the

East Timor operation. Japan initially sent a JSDF medical team under the disaster relief law, and then sent a 350-man engineering unit (of which 160 were support staff, not counted in the UN PKO figure) to the UN Stabilization Mission in Haiti (MINUSTAH) in February 2010 (still ongoing). The force's mandate was to clear the land allocated for camps for Haitians affected by the earthquake, repair roads and rebuild facilities. The JASDF was again involved in the transport of goods and personnel participating in the medical team from Florida to Haiti using a C-130. The JSDF deployment in Haiti was the first application of the current stand-by system, adopted in 2007 as part of institutional upgrading, as is detailed below, and this aspect of the operation proved highly successful.

In addition to UN-led operations, Japan's IPC law offered the primary normative provision for international relief operations outside UN mandate. In fact, under that law the JSDF provided assistance to Rwandan refugees in Zaire (now Democratic Republic of the Congo), to East Timorese refugees in Indonesia, Afghan refugees in Pakistan and to Iraqi refugees in Jordan.

The Disaster Relief Law was modified at the same time the IPC law was passed, paving the way for JSDF international relief operations. The paramount mission of the JSDF throughout most of its history had been domestic relief operations, which it conducted approximately 300 times a year on average. With the modification of the law, the JSDF conducted eleven foreign relief operations throughout the world, the most recent being the dispatch of a military medical team to Haiti.[11] Of these, particularly notable was the deployment of a joint JSDF force to Indonesia in 2005, following the earthquake off the coast of Sumatra and the tsunami that quickly followed. The 2005 case was the first joint JSDF operation under an integrated joint command, involving 230 JGSDF personnel, 3 CH-47s, 2 UH-60s, 640 JMSDF personnel, a tank landing ship (LST), JDS *Kunisaki*, the Helicopter Carrying Destroyer (DDH), JDS *Kurama*, and a fast combat support ship (AOE), JDS *Tokiwa*, as well as 90 JASDF personnel, 2 C-130s and 2 U-4s. In this case, the availability of a large amphibious platform proved essential to conduct the operation.

Japan's expeditionary missions have not been limited to UN PKOs and disaster relief, however, and Japan has adopted special measures

---

[11] These cases are: Honduras, Turkey, India, Iran, Thailand, Indonesia (mainly in Banda Ache), Russia, Pakistan, two more times to Indonesia and Haiti.

laws to allow the dispatch of JSDF for purposes other than PKOs and relief missions – for example, to serve on international anti-terrorism and stability missions. Later, Japan adopted a permanent Anti-Piracy Law. These have served significant and diverse security interests.

In autumn 2001, following American actions in response to the 9/11 attacks, the Anti-Terrorism Special Measures Law (ATSML) was passed in Japan after less than a month of debate in the Diet. Enabled by this law, the JMSDF was deployed to the Indian Ocean on a mission to refuel American ships on the high seas, and subsequently, warships of the countries of the coalition operating in the Indian Ocean, including the UK. The total cost of fuel provided to these countries amounted to a total of 21.9 billion yen (by the end of the operation). The mission's strategic importance was notable, because Japan was one of the few countries, other than the United States, that continuously maintained assets to provide fuel in that area. The ATSML was extended three times and was superseded by the Refuelling Assistance Special Measures Law in 2008. In January 2010, the latter law was not renewed and the operation was terminated.

Further, with the adoption in July 2003 of the Law Concerning Special Measures on Humanitarian and Reconstruction Assistance in Iraq (the so-called Iraq Special Measures Law), a 600-strong JGSDF unit was sent to Iraq.[12] The JGSDF mission operated between January 2004 and July 2006 in Samawah, in the province of Muthanna, located within the Multi-National Division (South-East) under British command. A total of 5,500 JGSDF personnel served on the ground. Their mandate was humanitarian and reconstruction assistance to the local community, primarily the provision of medical care and clean water, along with repair of roads and other public facilities. The JASDF was mandated by the same Special Measures Law to provide logistical support for transport of JSDF personnel and coalition troops, and humanitarian and reconstruction goods, and later UN humanitarian goods and personnel also, from Kuwait to Iraq. After the JGSDF withdrawal from Iraq, the JASDF mission was extended from Baghdad to Erbil in northern Iraq. The operation started in December 2003 and continued until September 2009, and accounted for the transport to Iraq of about 673 tons of humanitarian and reconstruction goods.

---

[12] The law was based upon a UN Security Council resolution (UNSCR 1483 – 2003) authorizing the Authority in Iraq, the humanitarian role to be played by the UN, and called upon Member States to provide humanitarian and reconstruction assistance to Iraq.

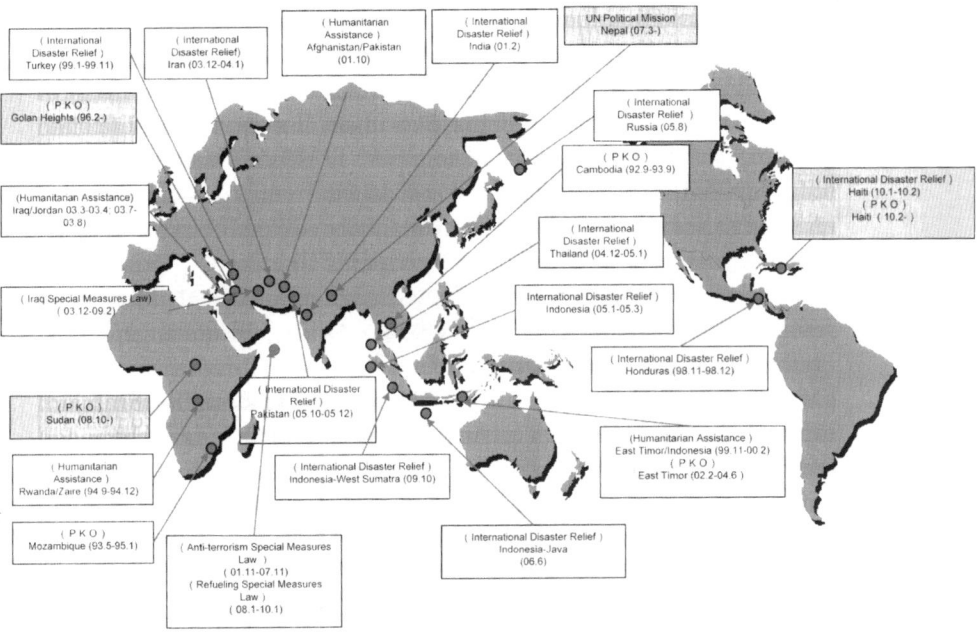

Figure 6.1. The JSDF in International Peace Cooperation
Source: JMOD (modified from original data) Reprinted with the permission of Ministry of Defense, Japan.

More recently, the JMSDF became involved in anti-piracy activities off the coast of Somalia and in the Gulf of Aden. Between March and July 2009, two vessels (JDS *Sazanami* and JDS *Samidare)*, each carrying SH60K patrol helicopters, special boarding units, motorboats and personnel for the Japan Coast Guard (JCG) were sent on maritime police activities, in accordance with Article 82 of the JSDF Law.[13] The new Anti-Piracy Law, which came into force in July 2009, is a permanent law, allowing the JMSDF to operate in any high seas and exclusive economic zones. Most notably, a fifty-member JGSDF team was also deployed to Djibouti for force protection missions for P-3C airfields, constituting the second joint mission that JSDF conducted since the disaster relief operation in Sumatra in 2005.

---

[13] Takai Susumu, 'Somalia Oki no Gendai Kaizoku Mondai no Hōteki Sokumen' (The Legal Aspects of the Current Piracy Problem off the Somali Coast), *Bōeihō Kenkyū*, Vol. 33, 2009, 10.

The number of engagements overseas to date suggests that, in contrast to the Cold War era, when expeditionary missions were virtually unthinkable, the post-Cold War era saw the emergence of expeditionary missions as a principal mission for the JSDF. This shift was accompanied by a redefinition of the elements of Japan's national security.

The second NDPO was issued in 1995, nearly twenty years after the first NDPO. Japan's role in preventing power vacuums and instabilities from occurring in the region was by then established, and the second NDPO, reflecting the radically changing post-Cold War security environment and citing diverse threats to peace, stipulated new roles for Japan and the JSDF in disaster management and international peace cooperation. The 1995 NDPO listed among the objectives of Japan's defence capability to 'contribute to the creation of a more stable security environment', one stipulation of which was to 'contribute to efforts for international peace through participation in international peace cooperation activities, and contribute to the promotion of international cooperation through participation in international disaster relief activities'.[14]

The subsequent defence review, the 2004 NDPG, further called for a more independent stance towards national security. National security was understood to rest not only on the ability to 'prevent any threat from reaching Japan' and if it does, to 'repel it and minimize any damage', but also on the ability to 'improve the international security environment so as to reduce the chances that any threat will reach Japan in the first place'.[15] The conduct of international peace cooperation by the JSDF was thus elevated to a primary position in Japan's new proactive and voluntary approach to cooperation with the international community.

The new definition of Japan's security role led to a set of institutional changes and upgrades. In 2007, the JDA was upgraded to ministry status, becoming the Ministry of Defence (JMoD). Also, international peace cooperation became a primary task of the JSDF, upgraded from the status of 'supplementary activities', in a move not unlike the American decision to give stability operations an equal status with combat operations. In 2006, the Joint Staff Council, in place since 1954,

---

[14] JDA, *The National Defence Programme Outline for FY1996 and Beyond* (Tokyo, 1995).

[15] National Institute for Defence Studies (NIDS), *East Asian Strategic Review 2008* (Tokyo: National Institute for Defence Studies, 2008), 222–3. See also, Japan Ministry of Defence, *National Defence Programme Guidelines, FY 2005* (Tokyo 2004), available at http://www.kantei.go.jp/foreign/policy/2004/1210taikou_e.html (accessed on 11 February 2011).

was replaced by the Joint Staff Office (JSO), which took over operational aspects and enabled a more integrated command.[16]

Another important institutional change was the creation of the Central Readiness Force (CRF). This component of the military had its own headquarters, and contained specialized units to deal with various contingencies, both domestic and international. Of particular relevance currently to international missions is the Central Readiness Regiment, under the command of the CRF. The CR Regiment has the capability to form advance missions to prepare for larger JGSDF contingents (and also responds to domestic contingencies) and is able to deploy thirty days in advance of main contingents. The CRF, furthermore, is the institution that organizes and commands all ongoing international JGSDF missions; and all future JGSDF missions abroad. The creation of CRF as a permanent command centre of all international missions the JGSDF conducts or participates in has had the effect of streamlining command and relieving the burden from (oft-inexperienced) regional armies and the Ground Staff Office (GSO).

As such, the CRF (its headquarters and regiment) forms a standby system, the first of its kind for Japan. In addition, Japan's five regional armies now take turns preparing 1,300-person stand-by units, from which suitable personnel with proper skill sets are selected for missions to be sent out within ninety days of the order. The recent dispatch of the JGSDF to the UN Stabilization Force in Haiti (MINUSTAH) is notable proof of the utility of such a stand-by system. The deployment of JGSDF units to the Haiti PKO was swift: within six days of the UN Security Council decision to expand the MINUSTAH force (UNSCR 1908), the Japanese government communicated to the UN Secretariat that it was prepared to send a JSDF unit to Haiti (25 January), and four days later, the UN Secretariat accepted the Japanese offer. The cabinet decision to dispatch JSDF to Haiti was made on 5 February and the next day, 160 of the soon-to-be 200-member advance party was sent from the CRF. The deployment of the advance party was completed on 14 February. The second dispatch, made up of the Fifth Brigade from the Northern Army based in Obihiro, Hokkaido (assigned a stand-by role under the rotation system) started on 24 February.[17]

---

[16] Japan Ministry of Defence (JMoD), *Nihon no Bōei: Heisei 21 Nenban Bōei Hakushō* (Defence of Japan 2009 – Tokyo: Japan Ministry of Defence, 2009), 153.

[17] Author's interview with JMoD senior official, and documents provided by the JMOD, 29 March 2010.

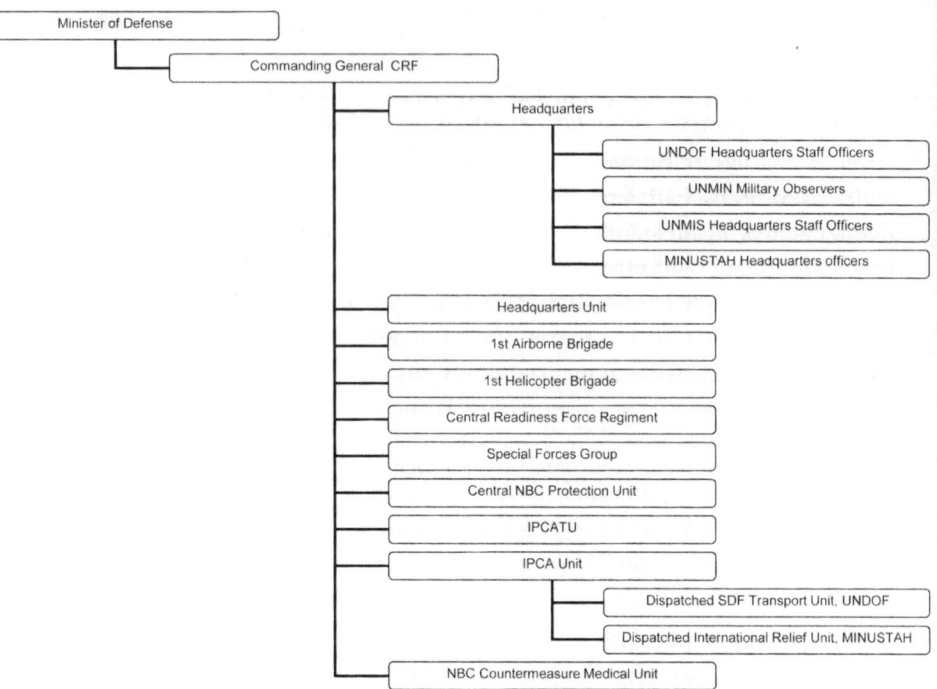

Figure 6.2. The CRF organisational structure
Source: Central Readiness Force (CRF)

Although it is not widely known, the JMSDF too has improved its readiness to deploy to international missions. The JMSDF underwent a major tactical reorganization in 2008, the salient feature being the restructuring of the four escort flotillas (the Cold War basic tactical formation) in eight escort divisions, each with four ships. These escort divisions are either helicopter-mounted destroyer (DDH) or guided missile destroyer (DDG) groups, depending upon whether the group is allocated a DDH or DDG. This structure replaced an older system wherein each of the four escort flotillas had eight ships, with one DDH as the flagship, and three escort divisions, each comprised of two or three ships. Previously, depending upon the mission, an *ad hoc* unit had to be formed each time a mission was assigned, and the relevant capabilities had to be patched together from among different fleets. The new structure allows the JMSDF to have one or two rapidly deployable fleets always ready, significantly enhancing its stand-by capability.

On the other hand, the JASDF had already created in 1989 an Air Support Command as part of its effort to strengthen national defence. This structure brought together tactical airlift, air-rescue wings, air traffic control, air weather service, and flight check groups. This new air command proved to be a helpful asset when the JASDF started to engage in international missions in 1992. In addition, the JASDF has sixteen C-130s, of which six are placed on stand-by for international disaster relief operations and are deployable within forty-eight hours.

Other relevant institutions to international missions include the CRF Airborne and Helicopter Brigades and the Special Forces Group, both trained to perform functions related to stabilization. Currently, however, these are intended for domestic missions and there is no legal framework that allows for their deployment abroad, although individual personnel from the Special Forces have been deployed to Iraq and to Haiti to provide security for designated persons. Should there be a political decision to deploy these capabilities, they will certainly constitute a significant asset.

In terms of education, within the CRF, an International Peace Cooperation Activities Training Unit (IPCATU) was created, devoted to the education and training of personnel who will serve in international operations. It not only offers basic training but also performs research aimed at enhancing education and equipment based on lessons learned, and it provides support of training for designated units of regional armies. The Japanese government has also decided to institute a separate International Peace Cooperation Centre (tentative title) to educate both military and civilian personnel. It is planned to be operative in 2011.

One further institutional upgrade that must be mentioned is the Japanese government's decision to join the UN Standby Arrangement System (UNSAS) in 2009. Japan is now registered under Level 1 in this system, which requires states to provide the UN Department of Peacekeeping Operations (DPKO) a list of types and numbers of personnel deployable to UN PKOs, the loosest of obligations among Levels 1-3. Under this agreement, Japan offers to send medical, transport, storage, communication, construction and machinery maintenance capabilities, if requested.[18]

In terms of equipment (or 'hardware'), the post-Cold War era has seen a greater focus by the JMoD on 'multi-functional' equipment to strengthen the scope/reach of its expeditionary capabilities. For example, in April 2010, 4 KC-767s (air-to-air refuelling aircraft) were domestically deployed, giving Japan that capability for the first time. The KC-767 is also a transport aircraft able to haul 30 tons. The range of

---

[18] As explained at http://www.mofa.go.jp/Mofaj/gaiko/pko/unsas.html (accessed on 11 February 2011).

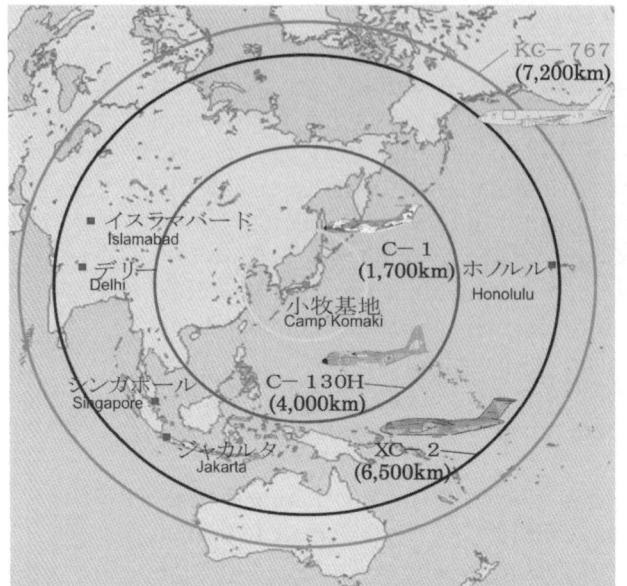

	Payload
C - 1	2 . 6 t
C-130H	5 . 0 t
XC-2	12 . 0 t
KC-767	30 . 0 t

Figure 6.3. Operational range of the JSDF air capabilities
Source: Ministry of Defense, Japan (Reprinted with the permission of JMOD from *Defense of Japan* 2009, p. 126.)

7,200 km (4,000 nautical miles, see Figure 6.3) of this aircraft takes the JSDF considerably beyond the geographical reach of the older versions of transport aircraft, such as the C-1.[19] The government is also working on developing the next generation transport aircraft, such as the C-X.

Another crucial step in the creation of an expeditionary capability concerns the JMSDF transportation platforms. In both the Iraq operation and the tsunami relief in 2005, the JMSDF deployed its 8,900-ton *Osumi* class of amphibious ships. These platforms are complemented by the JMSDF's four DDHs. Whilst originally designed for ASW functions, these are currently being replaced by newer models with enhanced capabilities suitable for expeditionary functions, the *Hyuga* class. In addition, it should be mentioned that the JMSDF mine-sweeping capabilities are among the best in the world.

Apart from these, Japan has a considerable number of helicopters that could be of use tactically in many stability contingencies. For example, the JGSDF has fifty-four CH-47s, and JASDF has sixteen. JGSDF has twenty-eight U-60s. The archipelago possesses also the option of outsourcing transportation for expeditionary missions to private operators, in addition to using the JSDF capabilities. Japan, as the

---

[19] JMoD, *Nihon no Bōei*, 126.

third largest economy in the world is able to contract out commercial ships and aircraft. For example, in the most recent mission to Haiti, Japan contracted *Antonovs* and commercial ships to transport equipment from Japan to Haiti.

In short, Japan has middleweight expeditionary capabilities set to increase with planned upgrading of certain equipment. This has come about in spite of the steadily diminishing defence budget (in absolute terms) over the past seven years and despite pressures arising from requirements for regional security and domestic relief missions. Expeditionary capabilities have been strengthened, in other words, without upsetting or transforming the basic force structure, largely through utilizing existing equipment and by adding multi-utility equipment that can be used in both domestic and international contingencies.

Operational and Doctrinal Limitations

Developments in material power did not correspond in adequate advancements in the realms of operational framework and doctrine. This is not to say that JSDF activities have no strengths – indeed, the JSDF's use of civil assistance is unique and has won the organization much acclaim to date. The JSDF have also high potential in what it brings to reconstruction projects, given the material resources that the country can spare. However, these advantages are often offset by the rigidity in the JSDF's missions, a particular feature the Japanese political system imposes upon missions abroad in regard to the nature of the deployment, mandate and authority.

Japan accepted the need to conduct expeditionary missions under some circumstances after the Cold War but within very constrained legal parameters. Since the passage in 1954 of the non-binding resolution prohibiting the dispatch of the JSDF abroad, Japan's military participation in UN missions abroad has come to depend upon whether or not the tasks to be performed involved the use of force.[20] Legal debates have dictated that Japan cannot constitutionally take part to UN-authorized multinational forces under UN charter Chapter VII (even if outside of the command of a multinational force) or in UN-commanded missions that involve enforcement (such as the second UN Mission in Somalia). The JSDF can, however, provide logistical

---

[20] Shibata and Soeya, 'Legal Framework', 51.

support to coalitions or multinational forces that engage in combat or enforcement, but only in areas that are not considered as part of the 'battlefields' and, under a separate command. To that end, JSDF deployments abroad are closely scrutinized to ascertain whether they would be required to operate in the 'battlefield'. All three branches, hence, were required to limit their operations to situations outside 'battlefields' in their past deployments, be they UN operations or Special Measures Law-related operations, such as the refuelling mission in the Indian Ocean or Iraq ground operations. Provision of transportation, excluding arms and ammunition, is permitted under the law. Actions that do not entail the use of force conducted in the same theatre as forces that use force or conduct enforcement are also allowed.

In addition to these legal considerations, there is an additional provision that limits JSDF activities. These are the so-called 'five principles' of PKO participation enshrined in the IPC law. The principles state the conditions against which the dispatch of Japanese personnel to UN PKOs is judged. The IPC law allows the dispatch of personnel only after a cabinet decision (without Diet approval, if the cases do not involve JSDF in infantry duties). By the IPC law, Japan may deploy personnel under the following five conditions: (1) a cease-fire agreement exists between/among warring parties; (2) there is consent from the state(s) and other parties to the conflict to allow Japanese personnel in the peacekeeping force; (3) impartiality of the force; (4) Japan retains the right to withdraw if any of the above conditions no longer prevail; (5) no use of force (or no use of weapons) except in self-defence.

These conditions are more akin to traditional peacekeeping than to contemporary peace-support or stability operations. Traditional peacekeeping relied on a trinity of principles: non-use of force except for self-defence, consent and impartiality. Contemporary peace operations, however, are characterized by both fragile cease-fire agreements and fluctuating levels of consent, which may require the peacekeepers to take certain measures of force to maintain credibility at the tactical level, if not at the strategic level.[21] Recent stability operations, furthermore, have departed from the 'trinity of principles' thinking. As is widely recognized in Japan as well, to strictly follow these five-principles would mean that there would be very few cases when Japan

---

[21] This is the understanding in the 2008 Capstone doctrine issued by the UN DPKO. Peace support operation in the NATO construct explicitly includes the category of 'peace enforcement'.

could in fact send personnel abroad. For example, these principles prevented Japan from offering logistical support to the operations of the International Force for East Timor (INTERFET), as requested by the UN. The fundamental problem remains that Japanese legal debates are not premised on the assumption that peace is fragile, peace processes sometimes break down, or that those who intervene may be faced with conditions of instability. It is evident that these five IPC conditions create serious complications in cases of complex stability missions involving hybrid threats.

Such legal frameworks also affected mandate and authority. The IPC Law listed sixteen specific tasks as IPC assignments, of which the first six concerned JSDF activities conducted either by units or individual members of the armed forces. These included the monitoring of cease-fire or the implementation of relocation programmes; the withdrawal or demobilization of armed forces; the patrol of buffer zones; the inspection or identification of weapons and/or their parts carried in or out by vehicles, other means of transportation, or passers-by; the collection, storage, or disposal of abandoned weapons and/or their parts; the assistance in the designation of cease-fire lines and other assimilated boundaries by the Parties to Armed Conflicts; and assistance in the exchange of prisoners-of-war among the parties to the conflict.[22] Other items on the list included: election monitoring; medical care; assistance and advice to police administration; distribution of food; repair of facilities, and so on. This constituted a 'positive list' of functions that JSDF units and IPC corps could carry out.

When the IPC law was passed, the Diet placed a temporary ban on JSDF units performing the first six tasks, effective until separate legislation was passed, although individual JSDF personnel could perform them (as happened in JSDF operations in Cambodia). In effect, this prohibited the JSDF from performing regular peacekeeping infantry duties unless there was a separate authorization from the Diet. This ban was subsequently lifted in 2001. The JSDF units nonetheless have not so far performed these duties, opting instead for logistical and humanitarian/reconstruction support.

The IPC law list of so-called core tasks cannot cover all the activities that the UN now routinely conducts in its operations, making it an inadequate framework for Japanese forces participating to UN PKOs. The IPC law list does not cover, for example, protecting freedom of

---

[22] International Peace Cooperation Law, Article 3(3) a-f.

movement; protecting UN facilities and personnel; and the protection of civilians, all of which are central areas of UN peace operations today. The JSDF are not allowed to be involved in a wider range of security-related tasks that the UN routinely conducts, such as recovering control of key facilities, assisting police and other security sector actors in the host nation, riot control, cordon and search operations, and enforced investigations, disarmament, arrest, and detention.[23] In the case of the JGSDF in particular, the grave implication of such a rigid methodology is that personnel on the ground are not always able to follow the force commander's orders, running the risk of creating vulnerability in the mission.

The JSDF often have been forced to devise ways to offset the impact of their own regulations using the means and authorities available to them. In Cambodia, for instance, after the UNTAC's mandate changed to emphasize security at election sites, JSDF personnel, with the ostensible objective of collecting and exchanging information, delivered water and food to UN Transitional Authority in Cambodia (UNTAC) personnel manning election sites so as to provide a measure of defence to these people by becoming a shield. The JSDF contended that had they come under fire, they would have been able to use weapons in self-defence, by extension offering de facto protection to UNTAC personnel.[24] This was a deliberate concatenation, as the JGSDF was not authorized to use weapons to protect other persons participating in UN missions.[25] In East Timor, faced with a UN order to provide security to UN facilities after the terrorist attacks in Bali in 2002, the JGSDF compensated for its inability to use force by fortifying the UN facilities with roadblocks and sandbags.[26]

---

[23] Michio Suda, 'Gendai Kokuren PKO no Setsuritsu/Unnei wo Meguru Seiji Rikigaku: Haiti PKO (MINUSTAH) wo Chushin ni' (Politics of Establishing and Managing Contemporary UN PKOs: The Case of the Haiti Operations – MINUSTAH) in Japanese Military History Association (JMHA, ed.), *PKO no Shiteki Kenshō* (Historical Reappraisal of PKOs – Tokyo: Kinseisha, 2007), 101–102.

[24] Tomoaki Murakami, 'Kanbojia PKO to Nihon: Heiwa no Teichaku Seisaku no Genkei' (Cambodia PKO and Japan: The Emergence of Peace Consolidation Policy) in JMHA, *PKO no Shiteki Kenshō*, 143–6.

[25] The 1998 amendment of the IPC Law made it possible for JGSDF personnel to use weapons to protect other staff in the UN mission, except military personnel. The latter prohibition of course is a reflection of Japan's ban on exercising the right of collective self-defence under the Japanese interpretation of the Constitution, an interpretation that erroneously applies the international law of self-defence to UN-authorized operations.

[26] Author's interview with senior JGSDF officer, 29 March 2010.

Another serious limitation concerns the gap between UN Rules of Engagement (ROE) and JSDF authority regarding the use of weapons. Japanese rules on use of weapons (i.e. ROE) do not cover all UN baseline ROE, consisting of ten items. The most notable exclusion in the Japanese rules is their failure to grant Japanese personnel the right to defend the mission, or the authority to detain, search and disarm to that end. Such actions are now a critical component in UN peace operations.

The authority of the Japanese forces operating under Special Measures laws has thus far proved to not differ from that under UN peace operations. For example, in the Iraq operation, where the situation was far more volatile than in normal UN PKOs, the authority of the JGSDF regarding use of weapons did not differ significantly from its authority in UN PKO contexts, even though some minor additional conditions were added to allow use of weapons.

Doctrine is another area where the JSDF have a degree of limitation in its action. Although the JSDF are gaining practical experience, there is no specific doctrine covering peace and stability operations (or none this author had access to it). On the other hand, from a review of past practice from the perspective of military doctrine, the following characteristics and principles of operations may be inferred.

(a) No use of force. Under the current legal system JSDF troops operating on the ground are able to use force in self-defence (defence of self and those under the control of the JSDF), but not for the defence of the mission, or for the defence of foreign military personnel, or the general population. This in effect amounts to no use of force at all, except for the purpose of self-defence, in a more narrow definition than that presented by the UN. As a result, in Japan's international missions there is a vast area that needs to be filled in regarding the spectrum of conflict between 'peace-time engagement' (as assumed in Japanese law regarding international missions) and all-out war. Currently, the JSDF do not have experience in using 'minimum force' 'necessary to accomplish a mission', since Japan refrains from use of force in international contexts.

(b) Further, the notion of prior consent is taken as an absolute in Japan's conduct of international peace cooperation, and this is a prerequisite for JSDF troop deployment in the context of the IPC law. As noted, the five principles of JSDF participation in UN

PKOs assume, in addition to the existence of a cease-fire, consent by the state(s) and other parties to the conflict to the participation of Japanese personnel in the peacekeeping force. This formulation correctly emphasizes the importance of consent in peace operations, but it fails to take into account the fact that consent in most contemporary operations is murky and fluctuating, and the fact that 'consent' as such is less and less frequently expected as a given at all times. In peace support operations, consent was understood to be a variable and therefore it could not be assumed; in contemporary stability missions, interventions have even more complex relations with local consent. In the Special Measures laws, consent was not required in the past, but hostilities from local parties were not implied, as the JSDF deploy only to areas that are not battlefields.[27]

(c)   Neutral, not impartial. As an extension of the above, neutrality, rather than impartiality, is what characterizes JSDF operations. The impartial implementation of a mandate, using minimum force if necessary to ensure a mission's credibility, is not recognized under the Japanese system/framework of operations.[28]

Taken altogether, the above principles stand firm into the traditional UN peacekeeping operations doctrine. Japan has yet to start thinking about filling the middle area, standing between traditional peacekeeping and peace enforcement, which was the focus of the international community during the 1990s. Officially, Japan does not assume the existence of this middle area, because the current interpretation is that the JSDF can deploy only to places that are not 'battlefields' (*sento chiiki*), and the five principles must apply in order for deployment as part of international peace cooperation activities to take place. In other words, Japan's international missions are always assumed to be 'peacetime engagement'. Such regulations do not generally take into

---

[27] In reality, the JSDF have used the 'hearts and minds' approach to build local consent. This was seen in East Timor, for example, and especially in Iraq, where consent building was understood as crucial to ensure force protection. There is, therefore, a significant disparity between what the JSDF deployed on the ground encountered or are likely to encounter and the basic legal presumption of conditions for deployment. A particular characteristic of the Japanese legal and political debate is the pretence that this disparity between rhetoric (legal rhetoric) and reality does not actually exist.

[28] In Japanese political debate JSDF activities are often referred to by citing the principle of 'impartiality', but this fact in itself indicates the insularity of Japan's discourse on principles of peace support.

full consideration the fact that contemporary conflicts rarely involve regular armies, clear-cut frontlines, or the *a priori* credible consent of local parties, and that a breakdown in peace agreements is always a real possibility. It is unclear what the JSDF could do if they were ever caught in such circumstances.

In addition, forces deploying in the context of peace operations must also have the skills to deal with riots. The JSDF train personnel in riot-control and policing, both as regular training and at CRF. There is a need, however, to make some further preparation in this area. For instance, Japan carried machine guns to Haiti (taking these to international missions often ruffles political sensitivities at home) but did not carry any non-lethal weapons, as requested by the UN, because the JSDF do not possess any.

There are some additional characteristics that are worth pointing out.

(d) Limited mandate. As discussed above, the JSDF mandate is limited, and as such is not designed to cover the full spectrum of contemporary conflicts. The JSDF mandate (see Figure 6.4) has so far excluded engagement in peacekeeping infantry duties under UN Charter Chapter VI as well as Chapter VII operations, especially enforcement. Notable exceptions are the JSDF participation in Chapter VII peace operations in East Timor and Haiti, utilizing their logistical and engineering capabilities. In these cases, JSDF missions included non-military activities such as medical assistance, provision of water and other civil assistance to non-coercive military duties.

(e) Civil assistance. Civil assistance is one of the JGSDF's strong points. This potential was recognized in East Timor, where JGSDF contingents were in charge of developing UN Civil-Military Affairs (CMA) projects. Further, civil assistance was the main focus in the JGSDF's Iraq operations. With the Ministry of Foreign Affairs also providing official development assistance (ODA) in Iraq, Iraq became an instance where civil-military cooperation took place in the manner of a provincial reconstruction team. The JGSDF also engaged in small-scale public relations and repair projects of roads and schools. The JGSDF civil assistance projects in Iraq used locally procured labour, providing employment opportunities to local Iraqis. This methodology was in contrast to that adopted in East Timor, where JGSDF personnel conducted much of the repair

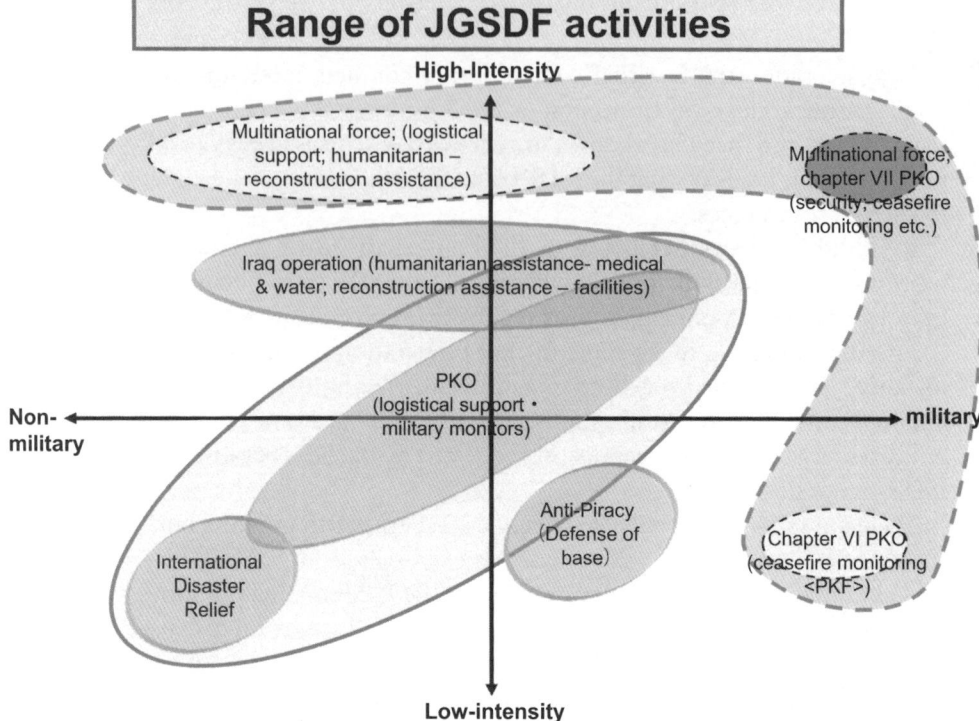

Figure 6.4. Range of the JGSDF activities
Source: Major General Koichiro Bansho, Annual meeting, Japan Society for Defense Studies, November 2009 (Reprinted with the permission of Major General Bansho).

and construction themselves. In addition, the JGSDF took an innovative 'democracy-promotion assistance' approach in Iraq. This sought to consult and coordinate local interests in deciding among the competing demands for reconstruction projects, thus ensuring both local ownership of reconstruction projects and winning over 'hearts and minds'.[29] The JGSDF's weakness regarding authority to use force served to advance civil assistance as a much needed force-protection measure in Iraq. Small

---

[29] The JGSDF found it difficult to balance these competing demands and do justice to all parties involved if they, the JGSDF, had to decide which projects to assist. They therefore asked the Iraqis to first go through their administrative chains to decide on the priority projects, in a democratic and transparent manner. The JGSDF would then check to assure that the local decisions had in fact been democratically made by monitoring both the demands and levels of satisfaction within the local community, as well as the requirements on the administrative side. For a more in-depth analysis of this topic, see Chiyuki Aoi, 'Beyond "Activism-Lite?": Issues in Japanese Participation in Peace Operations', *Journal of International Peacekeeping* Vol. 13, 2009, 72-100.

public relations projects and repair projects proved to be of vital importance in building good relations with local citizens.

(f)   Rebuild Initiative. The operational focus in past cases has clearly been infrastructure repair (rebuilding schools, roads, stadiums, electricity and water facilities, hospitals, etc.), and the JGSDF's engineering capabilities in clearing and building roads were heavily deployed in Cambodia and East Timor. When coupled with large ODA figures, this is an important strength of the JGSDF that might represent a template for Japan's future military contributions.

Overall, although there have been some distinct advantages to the commitment of the JSDF to expeditionary missions, there are still significant gaps in the relevant Japanese operational doctrine. The first and most obvious gap concerns the prohibitions on the use of force and the restrictive ROEs applied by the JSDF. This raises the question of whether they are really suited to today's operational environment, even in cases where the JSDF are asked to provide logistical support.

The second gap is the lack of a coherent stability concept in the Japanese doctrine. Because Japanese assistance is officially limited to 'post conflict' (as reflected in the five principles), stabilization becomes logically superfluous. Yet, from East Timor to Iraq, the JSDF have recognized the importance of civil assistance in stabilizing the situations.

Third, there is an important gap, conceptually and operationally, in the understanding of how Japanese activities can fit within the increasingly fuzzy fusion of concurrent activities that encompass various forms of irregular warfare. Specifically, it is not clear how JSDF activities may coexist, or relate to counterinsurgency. Officially, the restrictive Japanese law disqualifies Japan from participating in or being associated with counterinsurgency missions. Although the JSDF's potential to cater for civil assistance projects and engineering capabilities is considerable, Japan has so far ruled out ways to integrate these capabilities in a full-spectrum approach to counterinsurgency.

Finally, Japan's experience and doctrinal thinking in the area of joint operations remain relatively limited, despite the creation of the Joint Staff Office (JSO), with practical consequences for equipment acquisition and operational planning. More commitment at the JSO level to enhancing joint expeditionary capabilities may be warranted in order to allow for more effective and efficient equipment acquisition by reducing inter-service rivalry, and would also enhance operational planning.

Conclusions

As suggested at the outset of the chapter, an examination of Japan's experience in post-Cold War expeditionary missions, there is a significant gap between the country's material capabilities and the 'software' of its operations, the operational framework and doctrines. Unless policy-makers and the public decide to significantly update the latter, Japan's contribution to international peace and security will remain severely limited, allowing Japan only to 'punch below its weight'.

The fundamental question, however, is whether the Japanese people would find it in their interest to 'punch above their weight' in international security. The Japanese still hold their pacifist values close to their hearts, and they remain reluctant to become involved in conflicts abroad, especially when Japanese lives as well as the lives of local people are at risk. Senior politicians from realist circles, on the other hand, are not convinced that Japan, with powerful potential adversaries in the region, has any need to worry about anything beyond the regional balance of power. Thus low-cost 'international peace cooperation' based on obsolete principles and doctrine reflects the lowest common denominator for both pacifists and realists in Japan.

The priorities in Japan's security policy remain murky, and the debate over what those priorities should be to achieve security in an ever-more complex and competitive world is not growing more substantive or more constructive. What interests should guide Japan's expeditionary activities in various contexts and diverse geographical areas? How can Japan frame policy objectives and means, and formulate rationales on which to justify potential missions, and how should Japan direct its resources to achieve its policy objectives? These questions are still not being addressed. Japan's divided bureaucracies and weak political leadership under the previous and current governments have been liabilities in strategy making. In the meantime, demographic decline and slow economic growth, prompting reductions in defence spending, have had the additional consequence of forcing Japan to try to do more with less.

Nonetheless, this is a crucial time for Japan to develop strategy for dealing with the competing demands of the regional balance of power, expeditionary missions, and domestic contingencies. Japan's failure to develop such a strategy would affect regional stability as well as global stability. Japan has sufficient potential and capabilities at hand to become a positive force for global stability, should it so choose.

PART THREE

MARITIME STRATEGY IN AN INTERDEPENDENT WORLD

CHAPTER SEVEN

THE POLITICAL AND NORMATIVE CONSTRAINTS TO JAPAN'S NATIONAL SECURITY

Guibourg Delamotte

Looking back at the two decades that followed the end of the Cold War, it cannot pass unnoticed that Japan's defence doctrine regarding overseas deployments has changed. Yet this transformation notwithstanding, legal and political constraints remain considerable. The Constitution enacted in 1947 establishes in Article 9 that Japan renounces the use of force or threat thereof as a means to solve of international disputes, and that it shall not maintain any war potential for that purpose. From this constitutional provision a number of principles were derived. The interpretation of these principles led different post-war governments to create a series of 'self-imposed' constraints which in turn had (and continue to have) repercussions on Japan's military undertakings overseas and international commitments. The previous chapter addressed this question in relation to the realms of doctrines and operations.

This chapter takes the narrative into the realm of the Japanese debate on national security, examining the impact of the relationship between political action and legal norms on the country's defence policy. The chapter first reviews the genesis and development of legal constraints relating to overseas deployments, subsequently charting their initial relaxation throughout the 1990s. It shows how political power struggles throughout the post-Cold War period underpinned legal constraints. The result was that these limitations remained still significant and above all, that domestic politics heavily affected Japanese contributions to international military initiatives. Within this normative transformation, the chapter explores how the specific features of naval operations made the deployment of naval assets easier to authorize legally and less controversial politically, than land operations. In this respect, it analyses the political process that led to the legal measures authorizing the operations in the Indian Ocean and in the anti-piracy activities off the Somali coast. The chapter concludes with some final considerations on the impact of recent political changes in Japan on its security policy. In August 2009, the Democratic Party of Japan (DPJ) won the general election: how is Japan's defence doctrine likely to change with the DPJ?

The Political Foundations of Legal Constraints on International Operations

During the period from 1947, when the Constitution entered into force, to 1952, when the JSDF were created, the use of force was banned even for the purpose of self-defence.[1] When the JSDF were created, self-defence had to be admitted, but to guarantee that the new force would serve only self-defence purposes and would not make use of force to settle international disputes, their deployment overseas was banned. International missions were considered unconstitutional, regardless of their objectives.[2] This reading of Article 9 was set forth by a House of Councillors resolution of 2 June 1954 (*Jieitai no kaigaishutsudō wo nasazaru koto ni kansuru ketsugi*), adopted when the Self-Defence Forces (SDF) Law was passed.[3] Article 3 of the SDF Law indicated what the missions of the JSDF were; overseas deployments were not among them.

A first review of this notion appeared in the mid-1960s. The Cabinet Legislative Bureau (CLB) came up with the distinction between military deployment of aggressive nature (*kaigai hahei*) and a military deployment of peaceful nature (*kaigai haken*), to observe a cease-fire or upon UN request, for instance. According to the Chief of the Cabinet Legislation Bureau (CLB) Takatsuji Masami, it was conceivable for the JSDF to be allowed to take part in the latter provided certain conditions were met – for example, in case an entire unit was not deployed at once or the JSDF were needed for their military expertise.[4]

---

[1] Session 7, House of Representatives, plenary session N. 11, 23 January 1950.

[2] Cf.: Statement to the Budget Committee of the House of Representatives by Fujisaki Masato, Director of the Treaties Bureau of the Ministry of Foreign Affairs (MOFA), 26 March 1958. Yoshiharu Asano, *Kenpō Tōbenshū, 1947-1999* (Answers on the Japanese Constitution, 1947-1999 – Tokyo: Shinzansha, 2003), 159.

[3] Jieitai hō (SDF law), law N. 165, 9 June 1954.

[4] Session 63, House of Councillors, Budget committee N. 3, 3 March 1965.

This debate remained primarily at the theoretical level and bore no legislative consequence until the early 1990s. In this respect, it is worth pointing out that in 1987, a law allowing medical teams to be dispatched overseas was adopted, but JSDF members could not be part of those teams. The end of the Cold War and the Gulf War changed substantially the environment in which Japan's international cooperation took place; it could no longer contribute to US strategy and the protection of the 'free world' solely by defending its territorial boundaries. As the chapter in this volume authored by General Yamaguchi underlined, the shock of having contributed $13 billion to the liberation of Kuwait and receiving no gratitude caught political attention.

In October 1990, a first attempt was made at passing a law on cooperation with the United Nations (*Kokuren heiwa kyōryoku hōan*), creating special United Nations 'peace cooperation units' which could appear as institutionally distinct from the JSDF. In January 1991, the government considered amending the SDF Law's implementation decree to create a unit which could provide refugee assistance. Instead, it ended up sending a small flotilla of minesweepers to the Persian Gulf in April 1991, on the basis of Article 99 of the SDF Law. This provided that the Director of the then JDA could send the JMSDF on minesweeping duties to protect nationals.[5] A few years earlier, the CLB might have assumed that this provision was intended to be applied in territorial waters only, but times were changing. The government therefore justified the deployment by stating that there was 'for Japan as for the international community in general, a pressing need to restore security in the Persian Gulf'.[6]

In the aftermath of these events, a number of laws were passed, which allowed the JSDF to take part in United Nations peacekeeping operations (June 1992),[7] to assist the United States in 'situations in areas surrounding Japan' (May 1999),[8] to contribute to the international

---

[5] Hidetoshi Sotōka, Honda Masaru, Miura Toshiaki, *Nichibeidōmei no Hanseiki – Anzen to Mitsuyaku* (Half a Century of Japan-United States Alliance – Security and Secret Agreements – Tokyo: Naigai Shuppan, 2004), 431.

[6] 'Jieitai Hatsu no Kaigaihaken' (First deployment overseas of the SDF), *Asahi shimbun*, 25 April 1991.

[7] Kokusairengō Heiwa Ijikatsudō tō ni Taisuru Kyōryoku ni Kansuru Hōritsu (Law on cooperation to the United Nations Peace-keeping Activities), N. 79 of 29 June 1992.

[8] Shūheijitai ni saishite Wagakuni no Heiwa oyobi Anzen wo Kakuho suru tameno socho ni Kansuru Hōritsu (Law on the organization of the preservation of Japan's peace and security in relation with the situation of its environment), N. 60 of 28 May 1999.

fight against terrorism (November 2001)[9] and to be deployed in Iraq (July 2003).[10] Gradually, the category of deployment which was ruled out was narrowed down and the category of deployments deemed constitutional was broadened. The 1992 law admitted deployments with the consent of the state and the parties to a conflict, in the context of an international operation of the United Nations, to provide humanitarian aid. In 1999, the JSDF could be deployed to assist the United States when it intervened to ensure Japan's security or in the context of a regional crisis. In 2001, Japan could provide logistic support to members of an international coalition, in the vicinity of a third country which was in a civil war (which was not strictly speaking an international conflict). In 2003, the JSDF were deployed on the territory of a state which had no legitimate government yet, after a war, to help rebuild it alongside the members of an international coalition (though not as a member of it) – an 'occupation force' as the political opposition parties in Japan characterized it. Nonetheless, a core continuity with the spirit of the Constitution remained, and Prime minister Koizumi Jun'ichirō could therefore state in May 2001 that 'for the government, sending an armed contingent for the purpose of using force in the land, naval or aerial territory of another state ... exceeds the minimum level necessary for self-defence and is forbidden by the Constitution.'[11]

This statement notwithstanding, there remained practical implications to the fact that the JSDF were instituted for self-defence purposes. The first is that Japanese forces, once they are deployed, must comply with requirements which were established in 1969. At the time, CLB Chief Masami Takatsuji defined the three conditions of self-defence: damage must be imminent, there must be no other way to ensure Japan or JSDF member's defence, and use of force must be reduced to the minimal necessary level. He thereby distinguished reaction in

---

[9] Heisei Jūsannen Kugatsu Jūichinichi no Amerika Gasshūkoku ni oite Hassei shita Terorisuto ni yoru Kōgeki tō ni Taiō shite Okonawareta Kokusairengō Kenshō no Mokuteki tassei no tame no Shogaikoku no Katsudō ni taishite Wagakuni ga Jicchi suru sochi oyobi Kanrensuru Kokusairengō Ketsugi tō ni Motozuiku Jindōteki sochi ni Kansuru Tokubetsu sochi Hō (Special law on humanitarian measures based on a resolution of the United nations and on measures adopted upon the action of a number of states for the realisation of the objectives of the United Nations Charter, after the terrorist attack of 11 September 2001 in the United Stataes), N. 113 of 2 November 2001.
[10] Iraku ni okeru Jindō Fukkatsu shien Katsudō oyobi Anzenkakuho shien Katsudō no Jicchi ni Kansuru Tokubetsu sochi Hōritsu (Special law on the realization of activities of reconstruction aid and the restoration of order in Iraq), N. 137 of 1 August 2003.
[11] Session 151, House of Councillors, plenary session N. 23, 11 May 2001.

self-defence (*jieikōdōken*) from the use of force as a means of inter-national dispute settlement banned by the Constitution (*kōsenken*).[12]

As a result, when the deployment of the JSDF was authorized at the beginning of the 1990s, the use of force which the JSDF could exercise on the ground was strictly limited. Article 24 of the PKO Law allowed a member of the JSDF to use force when it was inevitable; retaliation had to be proportioned to the attack. He/she was allowed a small weapon to protect himself/herself or the JSDF members around him/her.[13] An order to fire was considered inconsistent with those requirements. Subsequently, the law was amended when the above-mentioned anti-terrorism law of 2001 was adopted: Article 12 of that law allowed the JSDF to use their weapons to protect people under their responsibility (*kanri*). That use of force remains particularly restrictive.

Limitations on the use of force extended further. The notion of 'asso-ciation with the use of force' was banned as well: the JSDF could not be part of an international force undertaking an armed operation, as this would constitute an exercise of force by proxy or 'fusion with armed violence' (*buryoku to ittaika suru*). But should not the purpose of the armed operation be taken into account?

The question of Japan's participation in an army operating for the UN was raised early on; in 1961, CLB Chief, Shūzō Hayashi, stated that Japan could participate to a UN army if this came to exist.[14] In December 2001, the CLB Chief, who at the time was Osamu Tsuno, took a different position on the subject, stating that participation in a multinational army was banned when implying being under the same command and acting as a member of that army, but some kind of cooperation with such an army might nevertheless be constitutional provided Japan was not associated with the use of force exercised by that army.[15] This interpretation led directly to crafting of the terms of the JSDF's Indian Ocean refuelling activities. In the mission, non-association with the use of force was guaranteed in theory by the fact that the destroyers were not directly firing at targets on land. That the operation was naval was a key aspect of its constitutionality – which to the eyes of the opposition parties remained questionable. In 2003, the

---

[12] Session 61, House of Councillors, Budget Committee N. 21, 31 March 1969.
[13] Session 123, House of Councillors, plenary session N. 22, 8 June 1992.
[14] Session 38, House of Representatives, Budget Committee N. 8, 10 February 1961.
[15] Session 153, House of Councillors, Committee for Diplomacy and Defence N. 12, 4 December 2001.

debate resurfaced in the context of the war in Iraq where the Japanese government, in order to deploy forces, never relinquished their command to the coalition forces.

An additional consequence of the ban on the use of force in overseas deployments is that these cannot take place in combat zones. Initially, when the PKO Law was adopted, the JSDF could not be deployed in countries at war. With the 1999 law on 'situations in areas surrounding Japan', rear support could be provided to the United States in a number of cases. In fact, 'rear support', or logistic support, was not a real novelty. It had been introduced for the first time in the 1961-66 Defence Equipment Objectives and mentioned subsequently in the 1978 Guidelines for Japan-US Defence Cooperation. In that context, 'the JSDF and US Forces will conduct efficient and appropriate logistic support activities in close cooperation in accordance with relevant agreements'. In 1996, an additional agreement was signed, the Japan-US Agreement on reciprocal provisions of logistic support, supplies and services (ACSA), establishing the nature of the support Japan could provide when the US intervenes in the event of an attack against Japan. Further, the new Japan-US Guidelines for Defence Cooperation, in 1997, stated that Japan could provide support to the US in the context of a 'situation in areas surrounding Japan'. Japan's rear support, '(b)y its very nature (was intended to) be provided primarily in Japanese territory' or on the high seas and international airspace around Japan as opposed to 'areas where combat operations (were) being conducted' (point V of the 1997 Guidelines).

Logistic or rear support was in effect the type of support provided in 2001 in the Indian Ocean where Japan provided refuelling aid and assistance to the wounded operating 'on Japanese territory, on the high seas or on the territory of a third country with the consent of the latter'. In Iraq (2003), where Japan contributed in humanitarian aid, reconstruction assistance and peacekeeping, it did so in 'non-combat zones'. Both laws indicate that Japan would conduct no act of war (sentō kōi), defined as 'murders and destructions occurring in an international conflict'.[16] An examination of the legal framework concerning the deployment of the JSDF overseas shows that legal measures evolved over time. Their transformation rested pre-eminently on the changing interpretation of the boundaries of the constitutional content.

---

[16] Katsutoshi Takanami, 'Heiwashugi no Genzai to Mirai' (Pacifism Today and Tomorrow), *Jurisuto*, N. 1222, 1-15 May 2002, 123-31.

Indeed, the nature and extent of legal constraints always stemmed from specific political circumstances.

The Transformation of the Political Debate in the 1990s

The first major political interventions to clarify the normative boundaries of Japan's defence doctrine occurred between the end of the 1950s and the second half of the 1960s. In particular, it was the government of Prime Minister Satō Eisaku, who was negotiating the return of Okinawa in the context of the Vietnam War, that sought to give the opposition a number of guarantees regarding the interpretation of the Constitution. The ban on arms exports was first worded in 1967 (at the time, it banned exports of weapons to countries from the communist bloc, involved in international wars or under international sanctions).[17] Prime Minister Satō also declared that Japan abided by three non-nuclear principles: it would not produce, possess or let nuclear weapons be introduced onto its territory.

In the early 1990s, both the political situation and the international context started to change. By 1996, the primary opposition party, the Socialist Party, had all but disappeared from the Lower House and, by 1998, from the Upper House. A generational shift was taking place inside the ruling party too, the Liberal Democratic Party (LDP). Longstanding leaders had been discredited by financial scandals; key figures like Hosokawa Morihiro, Ozawa Ichirō, Hata Tsutomu and others left the party, all causing the LDP's breakdown and prompting a systemic change. In particular within the LDP, there was a growing awareness that Japan should make a more visible international contribution given the new international context. When still in the party, it was Ozawa who, on 9 November 1990, reached a political agreement with Minshatō (a moderate-Left party) and Kōmeitō (related to a Buddhist sect), to see that a bill on cooperation to international operations was passed.

This agreement and Ozawa's determination were crucial in enabling the vote of the PKO law as the left stood by its original positions that the JSDF and the Japan-US treaty were unconstitutional. It is not until the Socialist Party found itself ruling in coalition with the LDP (from June 1994 to January 1996) that it admitted that the JSDF were

---

[17] Session 55, House of Representatives, Accounts Committee N. 5, 21 April 1967.

constitutional, though they should not be deployed overseas. Its positions regarding the US-Japan security alliance did not change: the security treaty should be replaced by a treaty of amity and the US Forces should leave Japan. By then, however, the Socialist Party did no longer count as an opposition force, as the main opposition party emerging was the DPJ, officially created in January 1996.

In the transition phase of the 1990s, the legal boundaries of Japanese military contributions overseas became a primary battlefield for Japan's political parties. In 1991, the impact of opposition parties on the PKO bill was centred on the insertion in it of five conditions in Article 1, also known as the five principles.[18] In 1999, the Bill on situations in areas surrounding Japan was voted at the same time as two others: the Bill amending the SDF Law and the Bill approving the ratification of ACSA. The LDP did not hold a majority of the Upper House; the Liberal Party and Kōmeitō, the second largest opposition party, voted for the three bills. The DPJ voted against the Bill on situations in areas surrounding Japan and for the other two. The Communists and Socialists were against all of them. In 2001, the DPJ initially voted against the Bill organizing the dispatch to the Indian Ocean. In the party's view, the Diet's approval was necessary before the dispatch could take place. This idea stood in contrast with the government's proposal for a dispatch approved by the government itself, with the Diet being informed of the decision. The other opposition parties were against the text as well, with Ozawa's new party voting against and drafting a Peace Cooperation bill in November 2001 instead.

A similar situation was recreated in 2003, when all opposition parties voted against the dispatch to Iraq. The arguments set forth were that Iraq was not pacified and the troops would have to use their weapons; their lives would be at risk. The Communist Party was appalled by the fact that Japan would occupy a country. The DPJ too, considered Iraq too dangerous; the JSDF would not be on a peace-keeping duty, but on a peace-making one. The DPJ's chairman of the time and later Prime Minister, Kan Naoto, called for a new resolution and for a force including France and Germany; only then, he insisted, should the JSDF go.[19] The DPJ asked for prior approval of the Diet to be inserted in the

---

[18] On this subject, see the chapter by Professor Aoi in this volume.
[19] Meeting with the United Nations Secretary General, 30 April 2004.

bill, for the mission to be brought down from four to two years, and for transport of weapons to be banned.[20]

The role of opposition parties was strengthened by internal political fractures within the LDP as well. These were centred on political figures who had been in the shadow of Prime Minister Koizumi, like former LDP secretary general Katō Kōichi and former chairman of the LDP's Policy council Kamei Shizuka, Kōmura Masahiko and Koga Makoto, both former contenders in the 2003 election for LDP leadership,[21] and Nonaka Hiromu, a member of the Hashimoto faction defeated by Koizumi in 2001. As a result, to win over the support of Kōmeitō, the government's coalition partner, Koizumi sent his president to Iraq in December 2003 and again one year later to ascertain the conditions of the deployment, before the renewal of the law. In all, in the aftermath of the Cold war, the transitional phase of Japanese politics meant that opposition parties have exercised constant political pressure on the government. They played a key role in limiting changes to Japanese defence doctrines and heavily influenced the nature of Japan's international undertakings.

Politics, Legal Norms and Japan's International Military Profile

One key aspect affected by Japan's legal constraints is the ability to deploy its forces in a way that they can be fully integrated into a multinational command and operational structure. The government's interpretation, spelt out in 1972, holds that Japan has the right to collective self-defence under international law (as the 1951 security treaty with the United States recognised) but cannot exercise it.[22] What this implies in practice is that the JSDF can benefit from another nation's armed forces' help, but cannot provide it in return. The limitations of such an interpretation became apparent especially in Iraq. There, the JSDF had to receive Dutch, Australian and British protection when around Samawah the situation had become unsafe but, had those militaries been attacked, the JSDF would have not been in the condition to fight back alongside. Similar limitations apply to Japan's missile defence system, which

---

[20] Tomohito Shinoda, *Reisengo no Nihongaikō* (Japanese Diplomacy after the Cold War – Tokyo: Mineruba Shobō, 2006), 99.

[21] 'Jikō datō na Handan' (Between Jimintō and Kōmeitō, an appropriate decision), *Yomiuri shimbun*, 10 December 2004.

[22] Session 69, House of Councillors, Budget Committee N. 5, 14 September 1972.

the country is jointly developing with the US. In theory, Japan could only strike down a missile headed towards Japanese territory. Reports were submitted to Prime Minister Abe Shinzō and Asō Tarō suggesting this interpretation be changed. The report addressed to the former and submitted to his successor, Prime Minister Fukuda Yasuo, identified several types of situations where admission of the right seemed particularly pressing (such as the ones described above).

The direct link between longstanding legal limitations and domestic political debates had the collateral effect on the ability of the JSDF to gain combat experience on the ground, but also to make them particularly weary of missions in hazardous environments. The lack of deployment to Sudan illustrates the JSDF's hypersensitivity to risk, caused by the fact that they feel their restricted right to use weapons can make them inadequate to the task. When Sudan's twenty-one-year civil war seemed to come to an end on 9 January 2005 after the government and rebel forces in the South signed a peace agreement, the Security Council authorized a peacekeeping mission to observe the truce and help the return of refugees. In a context where Japan was working on a joint-resolution proposal on Security Council reform with India, Brazil and Germany, Foreign Minister Machimura Nobutaka showed an immediate interest to deploy the JSDF.[23]

Three years later, Japan's interest in taking part to the mission was still in the making, with the Minister of Defence of the time, Ishiba Shigeru, stating that he was in favour of a dispatch there.[24] In May 2008, he announced that hundreds of civil engineers from the JGSDF might be sent for road construction and mine removal. By then, the situation seemed safer. Approximately 10,000 soldiers from seventy countries (including China and South Korea) had been dispatched as part of UNMIS. Eventually, two officers from the JGSDF were deployed in October 2008 for database management and transportation scheduling,[25] and their mission extended in June 2009. What made deployment to Sudan so difficult?

In 2005, the circumstances made a Japanese deployment a likely option. The JSDF could be asked to contribute to monitor the truce, which they could under the PKO law. Yet truce observation was considered dangerous and indeed, it was one of the JSDF's 'frozen missions'

---

[23] The 'G4' gave up on submitting such a proposal the following summer when it became clear they would not reach the desired majority.

[24] 'Ishiba OK with SDF dispatch to Sudan', *Kyōdo News*, 17 May 2008.

[25] 'GSDF officers to help UN rebuild Sudan', *Kyōdo News*, 28 September 2009.

until 2001. General Mori Tsutomu, Chief of Staff of the JGSDF, stressed that it would be difficult for them to participate in main peacekeeping operation activities unless troops were allowed to use their weapons for more than self-defence to accomplish their missions.[26] An operation conducted in Africa between September and December 1994 had left bad memories. On what was Japan's first humanitarian operation, 400 members of the JSDF provided medical assistance, sanitation, water supply and airlifts in Zaire (Goma) and Kenya (Nairobi) to the million refugees who had fled from Rwanda. There was a sense that the Sudanese situation might be similarly inextricable for peacekeepers.

Concurrently, political pressure for an intervention in Sudan was relatively weak (unlike after 9/11 and for Iraq). There was some political interest from the Ministry of Foreign Affairs in relation to the bid that Japan, Germany, Brazil and India sought to submit to the UN General Assembly. This concerned the proposal for a resolution on their joint candidacy as permanent Security Council members. Yet, at that time, Japan was engaged in Iraq and the situation in Sudan was dangerous. The G4 attempt failed in August 2005 due to lack of support from the African Union. Prime Minister Koizumi was at the end of his third mandate and had already led a political battle over deployment to Iraq. Subsequent Prime Ministers Abe Shinzō (until September 2007), Fukuda Yasuo (until September 2008) and Asō Tarō (until September 2009) had more complex political arenas to face, as they governed under a 'divided' Diet, with a DPJ in possession of the majority of the Upper House in the July 2007 election. In 2008, Prime Minister Fukuda gave some consideration to Sudan when he hosted the Fourth Tokyo International Conference on African Development (TICAD) Conference, which is why his Minister sought to send troops there, but the JSDF were no longer needed: the international operation had long been in place. Hence, the belated dispatch of a small number of officers only.

Nonetheless, between May 2008 when Ishiba Shigeru stated hundreds of engineers would be sent, and October when two officers effectively went, six months elapsed and nothing remained of the initial statement; the JSDF were able to delay until the government and the situation had changed, and the dispatch was neither called for by the government, nor needed by the international community. The lack of a deployment to Sudan shortly after the May statement raises

---

[26] 'Diplomats playing politics with PKO', *Yomiuri Shimbun*, 11 February 2005.

important questions on the need for the government to determine standards by which to assess whether or not participation in an intervention is in Japan's national interest so that deployments can take place swiftly. Furthermore, even with the JMoD favourable to it, this example would suggest that the current approach to Japan's normative limitations can become a tool for the military commands to express negative opinions on a deployment. Crucial to the JSDF's reluctance was the fact that they would not be able to defend themselves adequately.

One final aspect of Japan's international military profile affected by the politics of legal constraints is the actual duration of overseas missions. As this chapter has thus far shown, parliamentary control over operations is exerted in several ways. First, operations have been limited to the provision of humanitarian and reconstruction aid, or logistic support. Second, laws often included an obligation to seek the Diet's approval for the deployment plan and to inform the Diet of the conduct of the operation. The third element concerns the limited duration of operations, authorized always on the basis of special laws – such as the 2001 and 2003 laws. This option has implied that the Diet retained a significant power over the mission itself. The case of the Indian Ocean deployment under the 2001 Special Measures Law is a case in point. Initially, it provided for a two-year deployment following which the government had to seek a new law for a new maximum of two years. The law was renewed in October 2003 for two years, in October 2005 for a year and in October 2006 for another year. By 2007, as the political situation had changed, consensus over the mission did not exist and the deployment was no longer renewed. Only when a new law was drafted in January 2008 did the mission resume. That law was renewed once by the Asō Cabinet in January 2009 and expired in January 2010.

Reconciling Politics and Legal Constraints:
The Specificities of Naval Operations

Though Japan's legal constraints on the use of force have been (and continue to be) a highly politicized matter, one area of military operations seems to have represented a terrain where politics and legal issues are more easily reconciled. Indeed, naval matters pioneered in many respects changes in Japanese legal restrictions. During the Cold War, the defence of sea lanes prompted the first transformation. At the time, alliance management issues were of the utmost priority in the government's agenda, and this was one of the areas of Japan's defence policy where by defending waters adjacent to the archipelago Japan contributed to the United States' Asian strategy.[27] In May 1981, this was a core reason that enabled Prime Minister Suzuki Zenkō to pledge to a 1,000 nautical miles limit for the area that could be patrolled by the JMSDF. This was an impressive commitment since the 1976 NDPO had merely determined Japan's commitment to the surveillance of neighbouring seas. In particular, from a legal perspective, the 1,000 nautical miles limit went well beyond the 200 nautical miles exclusive economic zone in which under the 1982 Montego Bay Convention of the Law of the Sea coastal states exercised 'residual rights'.

In this respect, the internationalization of maritime operations started before the end of the Cold War and since the 1991 deployment of minesweepers to the Persian Gulf they paved the way in expanding the reach and scope of Japan's military operations. The most significant have taken place recently, illustrating the part they play in the internationalization of JSDF missions. In the aftermath of 9/11, the JMSDF were the first to be mobilized to provide support in the Indian Ocean to the Combined Task Force-150 (CTF-150) in the context of the Maritime Interdiction Operation side of Operation Enduring Freedom (OEF-MIO), an international coalition formed on the basis of Security Council resolution 1373 (2001). Initially, the US received the bulk of Japan's assistance and the share of provisions to other countries was residual. This gradually changed and the US fell below 50% of the total support offered by Japan's refuelling ships in 2005. The refuelling mission, whilst opposed in left-wing political circles, enabled Japan to show support to its primary ally from the onset of the military operations. Militarily it was low risk, and therefore politically acceptable. Strategically, the JMSDF's contribution was particularly welcome as Japan possessed excellent fast combat support ships, whilst it overall limited Japan's economic burden. By comparison with other countries anti-terrorism military budgets, the cost of Japan's economic assistance in the form of refuelling remained relatively small; in 2008, the figures were $13 billion for the US, $1.5 billion for the UK, $850 million for Germany and $70 million for Japan.[28]

---

[27] For a comprehensive treatment of this topic, cf. Graham Euan, Japan's Sea Lane Security, 1940-2004. A Matter of Life and Death? (Nissan Institute/Routledge Japanese Studies Series, London and New York: Routledge, 2006).

[28] Naikakukanbō, JMOD, MOFA, *Bōei Mondai Seminā: Tero ni Tachimukau Nihon, 9.11 Tero kara Nananen, Nihon no Yakuwari wa?* (http://www.mod.go.jp/rdb/kyushu/seminar/006/006-01.pdf, accessed on 10 May 2010).

The JMSDF has also been the first armed service to be involved in homeland security type of missions, conducted in support of the Japan Coast Guard (JCG). In March 1999, they pursued two spy-ships and fired warning shots until they lost them: it was the first time that they were asked for back-up by the JCG. In October 2001, Articles 20 of the JCG law and 91.2 of the SDF law were amended to allow the JCG and JMSDF vessels to fire directly on suspicious ships (prior to the amendment, only warning shots were allowed). In November 2004, the JMSDF chased a Chinese submarine until it left Japanese waters. Timing was interesting in both instances: in 1999, the Obuchi government was about to put to vote the three 1999 laws in connection with situations in areas surrounding Japan; in 2004, the Koizumi government was about to publish new defence guidelines. Intrusions occur often: not all receive as much attention.

The most recent naval deployment occurred off the coast of Somalia. On 14 March 2009, two destroyers were sent to the Gulf of Aden to protect Japanese vessels or vessels transporting Japanese goods or having Japanese nationals aboard. The naval group was dispatched on the basis of Article 82 of the SDF law, the basis for the JSDF's duties in relation to the maintenance of public order. Eight members of the JCG were with them. This was the JMSDF's first police operation outside Japan's territorial waters; the JCG had been in charge of the fight against piracy in the Malacca Strait since the 1990s. The law voted in June 2009 had a wide scope, allowing civilian ships to benefit from protection whether or not they are Japanese or transport Japanese nationals or goods.[29] For the first time, a law provided that the JMSDF may engage a hostile force (pirates in this case) in order to accomplish its mission.

Whilst leading the way to new missions abroad, naval interventions did not escape the political battles that distinguished Japanese international contributions. In the case of the anti-piracy mission, the process that led to the adoption of the measures to dispatch the naval group to the Gulf of Aden was characterized by the DPJ's intention to make its majority in the Upper House felt. In December 2008, Ozawa Ichirō, the party's leading figure, declared that there was no doubt that the Japanese Constitution allowed Japan to defend its ships when they were threatened by an act of piracy. Yet in January, he further added that

---

[29] Kaizoku Kōi no Shobatsu oyobi Kaizoku Kōi no taisho ni Kansuru Hōritsu (Law for the repression and fight of acts of piracy), N. 55 of 24 juin 2009.

'nationals or Japanese goods had to be protected, but whether or not to send the JSDF to do so was another matter'.[30] By mid-March, when the first ships were to set sail towards the Somali coast, he had expressed concerns on the use of weapons that should be allowed (the JSDF were to be allowed to scare off the pirates 'if they did not obey the order to freeze and approached dangerously').[31]

Significantly, the DPJ insisted that the law made the JCG responsible for the fight: only if they struggled were the JMSDF to be called in. The DPJ called for the government to ensure better cooperation between the two forces and that action be taken simultaneously to improve the stability of neighbouring countries. The party also requested that the government sought prior approval from the Diet submitting a deployment plan including details of the envisaged patrol area and length of the mission. It further asked for any extended right to use weapons to not include assets belonging to other state actors and that the law should establish ground rules on how prisoners, if any, should be dealt with.[32] The DPJ rejected the government's bill in June 2009, which eventually was passed owing to the LPD's two thirds majority in the Lower House.

Where does this leave the question of Japan's legal boundaries in conducting military missions overseas? Japan's defence policy is the outcome of a sixty-year evolution in the interpretation of the Constitution. As the examination presented thus far demonstrated, legal constraints no longer serve their original purpose – to protect pacifism and democracy. Pacifism no longer means what it meant in 1947, it has been adjusted to match the changing international and national contexts. However, since these changes did not lead to an open debate on constitutional issues, relaxation on past constraints took place very gradually. New rules were approved under 'special' and 'limited' conditions. This would lead to the consideration that where and under what circumstances Japan is likely to deploy troops depends on political factors, the charisma of a Prime Minister, agreement of the ministries of Foreign Affairs and of Defence and the Prime Minister's office, or the weight and cohesion of a political majority.

---

[30] 'Minshu sanpi Shimesezu' (The DPJ does not agree), *Yomiuri shimbun*, 29 January 2009.

[31] 'Sentai Shageki Kitei Kokkai Shōten ni' (Provision on shots focus of Diet's attention), *Yomiuri shimbun*, 15 March 2009.

[32] 'Kaizoku taisaku, Kaihō shutai ni' (JCG to be central in anti-piracy policy), *Yomiuri shimbun*, 20 March 2009.

The experience of the past two decades suggests also that the political use of legal constraints limits further change rather than to direct it. The senior military leadership lacks a sense of what it needs to report to its civilian and political counterparts. Prohibited actions are so numerous and any measure so sensitive that young officers fail to realize the particular importance of protecting some defence-related information. Furthermore, from 1952 the country's military posture has been based on the premise that self-defence was admissible. But would armed forces with no combat experience aptly defend their country? The peculiar nature of naval operations, happening far from public attention, in places where strategic and political impact can be high and risk of danger and costs comparatively low, offered significant advantages. They enabled the Japanese governments to find, in the majority of cases, adequate solutions to meet national security requirements and increased expectations on the country's international military contributions. In many ways, they helped Japan to move into the uncharted waters of international security, paving the way for the progressive relaxation of existing legal limitations. Yet, naval operations could not provide a comprehensive answer to what remains today a problem held hostage of domestic political power struggles.

The DPJ in Power: A New Course?

The general election of 30 August 2009 set the conditions for a new course in Japanese politics. The DPJ gained a majority in the Lower House and was elected by the Japanese people on a mandate for change. With respect to foreign and defence policy, the party campaigned on the grounds that it would behave differently from the LDP, a promise to make also sure that the party's coalition partners, the Social Democrats (six seats) and the New People's Party would remain satisfied. The DPJ, however, represented a collection of smaller groups with very different views on core security issues, and it cautiously avoided exposing divergences between its members in the year before the election. The party therefore appeared as further to the Left and as more homogeneous than it really is.

Indeed, at the time when the party took power in September 2009 it was made up of numerous groups which quite resembled the LDP's factions in structure. The most influential were headed by Hatoyama Yukio, Ozawa Ichirō, Maehara Seiji, Kan Naoto and Hata Tsutomu. On defence policy, the DPJ's group structure gauged divisions similar to those that split Japan's political scene during the Cold War. It combined former left-wing party members, who seceded from their former parties because they were more liberal (inclined to admit the Japan-US Alliance, for instance) and former LDP ones (leaving the party to promote reforms). In this context, as early as 24 February 2009, Ozawa, then the Head of the DPJ, declared in a press conference only one week after meeting Secretary of State Hillary Clinton that if Japan was to play a more important role in regional security, the US's role would decline and there would be no point, then, in the 7th Fleet remaining in Japan.[33] He hurried to add, three days later, that Japan would never claim the legitimacy to intervene in other nations' internal affairs and he had not meant to imply that Japan should strengthen its military potential.[34]

A few months later, just before taking up the role of first DPJ Prime Minister, Hatoyama Yukio, published a paper (partially translated in the *New York Times*) that gave the impression of a degree of Anti-Americanism in his political approach. In fact, the party platform in 2009 wanted for Japan a more pro-Asian position, calling in the meantime for an 'equal' and 'close' partnership with Washington. The party sought for Japan an autonomous foreign policy, assuming its share of responsibilities in the context of the Alliance, whilst admitting the idea of 'split responsibilities', of different duties for the two countries. This had been the party's stance since 2006.[35]

One core issue that the DPJ indicated the intention to tackle concerned the Status of Forces Agreement between Japan and the United States.[36] The main point to address cregarded the relocation of Futenma, addressed in a series of high official talks which took place at the beginning of Hatoyama's term. Whilst an agreement was reached in 1996 for the reversal of the territory occupied by Futenma, in May 2006, it was agreed that replacement facilities would be built further North, near Camp Schwab (on Cape Henoko, city of Nago). In its *Okinawa Vision 2008*, the DPJ question the agreement, stating that Futenma should be relocated outside Okinawa, or elsewhere provided

---

[33] 'Tōtotsu hatsugen hihan no uzu' [An abrupt declaration creates a swirl of criticism], *Yomiuri shimbun*, 27 February 2009.

[34] 'Bōeiryoku no kyōka hitei' [Reinforcement of defence potential denied], *Yomiuri shimbun*, 28 February 2009.

[35] Seiken seisaku no kihon hōshin (Seiken maguna karuta), 2006 [Fundamental policies, Minshutō's Magna Carta for government].

[36] Minshutō seisaku index 2008 [Index of Minshutō's policies, 2008]

the strategic environment allowed.[37] Since then, little real progress has been made, and the weak political position of the DPJ, with Hatoyama resigning from power a mere ten months after taking power, prevented the DPJ developing further its security policies. By the end of 2010, a new government formed by Kan Naoto showed to differ little in its assessment of the Alliance's significance to Japan's security and in its appraisal of the Alliance as a pillar of Japan's foreign and defence policy.

Conclusions

Almost sixty years after the establishment of the JSDF, political parties in Japan still debate their functions and contribution to the country's national security and foreign policies. As this chapter has shown, this debate made of the interpretations of the legal boundaries of the Constitution a primary battlefield in the defence and security policy-making process. The DPJ's recent rise to power does not seem to have put in place the mechanisms to change this, due to very different opinions internally. Its more radical members, a group called *Sōken kaigi*, published in 2005 a declaration introducing the notion of 'peace-creating nation', due to replace the current 'peace-receiving' Japan.[38] The document proposed the creation of a special unit for overseas interventions (so as to keep the main body of the JSDF within Japan's territory). The Constitution would also establish the limits of self-defence, the definition of which would be aligned with that of the UN Charter. In the document, the distinction between individual and collective self-defence is dropped (a position restated in the *2008 Policy Index*). The use of force is limited to the minimum necessary level. A law on security, annexed to the Constitution, would explain Japan's fundamental defence posture, the principles guiding actions in a state of emergency and civilian control, as well as the organization of the special UN unit.

The content of the 2005 *Sōken kaigi* report shows the extent to which political debate on defence issues evolved. The DPJ – in contrast to left-wing positions of the past – acknowledges the current role of the SDF and seeks to contribute more actively to international peace and to admit the exercise of collective self-defence. The DPJ, however, ties the

---

[37] *Okinawa Bijon 2008* [Okinawa Vision 2008].
[38] *Kenpō mondai gakushû shiryōshū* (Documents on Constitutional Issues – Tokyo: Gakushū no Yūsha, 2006), 29.

question of overseas deployments to a UN mandate. In legal terms, this idea underpinned the DPJ's approach to the law on the refuelling mission in the Indian Ocean when it expired in November 2007. Ozawa, the DPJ's leader at the time, stated then that Japan should take part in the International Security Assistance Force (ISAF, mandated by the UN and operated by NATO) and provide food and medical assistance, assistance in collecting weapons and structural consolidation expertise.[39] On 21 December 2007, the DPJ proposed a bill to authorize such a mission. It contained important legal innovations: the right for the JSDF to use their weapons to lift an obstacle in the completion of their mission; also, should an incident occur, they would not be subject to common criminal law.[40]

Contrary to commonly-held views, the DPJ is not against a more autonomous, more equal Japan. Past concerns expressed on the legal constraints in overseas deployments were more the result of domestic political power struggles than of an overall opposition to an international military profile. The extent to which further change will take place is difficult to predict, and the DPJ's weakened political stance is unlikely to foster more permanent changes in the norms regulating the missions of the JSDF. Special measure are likely to remain the main tool to reconcile different political views. For similar reasons, naval deployments are likely to remain the prime feature in Japan's overseas commitments, enabling the governments of the day to meet complex domestic and national security agendas. In this respect, it is worth noting that in the preliminary stages of the measures adopted to send the naval group off the Somali coast, one of the first options considered by the JMoD consisted of a refuelling operation modelled after the Indian Ocean's.[41] At sea, Japan's political divisions might find shape and develop the legal boundaries of the country's international military profile.

---

[39] Ichirō Ozawa, 'Ima Kōso Kokusai Anzenhoshō no Gensoku Kakuritsu wo' (For the immediate establishment of principles on international security), *Sekai*, 2007:11, 148-53.

[40] 'Minshutō Shintero Taisaku Hōde Danzoku Shingi Senjutsu' (A strategy of continued debate on the DPJ's new anti-terrorism propositions), *Sankei shimbun*, 21 December 2007.

[41] 'Somariachū Ten'yō "Mutdukashii"' (Move to the Gulf of Somalia thought 'difficult'), *Yomiuri Shimbun*, 28 October 2009.

CHAPTER EIGHT

'BACK TO AN OFFSHORE FUTURE': THE ROLE OF THE PAST IN BRITAIN'S CONTEMPORARY DEFENCE POLICY

Steven Jermy

It was clear throughout most of the twentieth century which state actors threatened democracies in Europe and the West, a fact highlighted in the chapters of this volume. Defence policy thus had a binary clarity. Defence was about defending your territory and securing the national interests abroad. And defence policy was thus about how a government would pursue these two aims. The defence policy planner's approach was straightforward. Design, build and configure your forces to defend yourself against your strongest likely enemies. And then adapt, from within this force structure, for anything else that you might want to use military power to deliver. The totality played out in three World Wars – two hot and one cold – and the policies chosen by the West prevailed, generally. At least this is what the historical record appears to show.

But events and the uncertain and changing nature of contemporary threats have complicated the position in the post-Cold War era. Defence policy has become a rather amorphous subject. In this modern world, defence for the West is about national security, rather than national survival. And because national security is an altogether more difficult subject to get your head around, then the design challenge for defence policy planners is more acute than ever before. It is not that we are without experience to draw upon. In stark contrast to the Cold War, the armed forces of the West have seen an unprecedented high level of operational activity. In the British experience, the UK has attempted open-heart surgery on nation states as far apart as Bosnia, Kosovo, Sierra Leone, Iraq and Afghanistan. And with somewhat mixed results – for both patients and surgeons.

But for all this new complexity, these post-Cold War operations share a common characteristic. Put simply, they have been 'interventionist'. Although the West's overarching political objectives have been defensive in nature, strategically they have been pursued with an offensive interventionist approach. This is an approach that for some nations – most obviously maritime nations such as the United States and Britain – runs counter to their historical preference.

As Sir Julian Corbett, Britain's most accomplished strategic thinker, noted in the early twentieth century, maritime states – particularly island nations – have a geopolitical advantage when compared to those on continents – their territorial integrity can only be threatened by sea or air. So, as long as they retain strong navies and air forces, they have no need to maintain large standing armies. It is for this reason that until the First World War the standing British and American armies were, in comparison to those of Europe and Asia, tiny and inexpensive. Both countries averred an interventionist approach, tended instead to maritime defence policies, banked the small-army dividend, and reaped the economic rewards unfolding from such an approach.

In this Chapter, the defence policy lessons of recent interventions are assessed against the successful maritime policies of the nineteenth and early twentieth centuries as a way to glance beyond Afghanistan, the UK's current largest military commitment overseas. The analysis is divided in five stages. First, the chapter will review Britain's post-Cold War defence policy, as declared. Second, it will consider Britain's post-Cold War defence policy, as practised. Third, it will examine the aggregate impact that British defence policy, both as declared and as practised, has had on Britain's national security. Fourth, it will consider the modern strategic context and examine the factors that now bear on Britain's national security, including current military operations in Afghanistan, broader threats from Islamic terrorists, and other plausible challenges for British defence policy in the decades ahead. Finally, it will set out the key principles underpinning an 'offshore' defence policy, and outline the core capabilities needed to implement such a policy.

British Post-Cold War Defence Policy – As Declared

In the two decades that followed the Cold War, British defence policy was revised on four separate occasions. Notwithstanding their political packaging, the first two White Papers – *Options for Change* in 1990 and *Front Line First* in 1994 – were savings exercises. They sought to bank the peace dividend. The second two White Papers – the *Strategic Defence Review* in 1998 and *Delivering Security in a Changing World* in

2004 – were more considered. The first was explicitly foreign policy led. The second sought to first make sense of, and then respond to, the Twin Towers attacks. The nature and shape of Britain's armed forces is, in very large part, a consequence of the judgements and decisions made in these four White Papers.

In terms of both force structure and expenditure, *Options for Change* in 1990 set the armed forces on a downward trajectory. Service personnel cuts amounted to nearly 20%. And this trend would continue in 1994 with *Front Line First* (or the *Defence Costs Study*). Although purportedly targeted at tail rather than teeth, *Front Line First* further reduced service personnel by 5% and would keep defence expenditure as a percentage of GDP heading south. Thus, by the time the Blair Government took power in 1997 service personnel had reduced by nearly a quarter; defence expenditure as a percentage of GDP had fallen below 3%, having been over 5% just a decade before, in 1986. Of course, with the collapse of the Soviet Union, such reductions seemed to make sense. But this view rested on the key assumption that underpinned both reviews – that there would be a reduction in threats to UK's security. For a peace dividend to be safely realized, there would need to be peace.

The *Strategic Defence Review* (SDR) of 1998 was the most important White Paper of the post-Cold War period. Anticipatory work began well in advance of the 1997 election and by 1998 the policy was based on very robust analysis. The SDR recognized, upfront, the new security context:

> (…) reflecting a changing world, in which the confrontation of the Cold War has been replaced by a complex mixture of uncertainty and insta-bility. These problems pose a real threat to our security, whether in the Balkans, the Middle East or in some troublespot yet to ignite. If we are to discharge our international responsibilities in such areas, we must retain the power to act. Our Armed Forces are Britain's insurance against a huge variety of risks.[1]

This, then, was a fairly broad canvass on which to work. What it perhaps unconsciously reflected, more than anything else, was the difficulty posed to defence planners in the amorphous idea of national secu-rity in the post-Cold War world. For example, how exactly did insta-bility in the Balkans prejudice the security of the British homeland? As we reflect, a decade later, on operations in Afghanistan, this dilemma challenges us still.

---

[1] UK MoD, *The Strategic Defence Review* para 2.

The SDR nevertheless established key design principles. A joint approach, with all three services working more closely together, was championed. Expeditionary operations were emphasized. A premium was placed on the British armed forces ability to react quickly, and on the inter-operability that would allow the services to act in concert with multi-national institutions, particularly NATO. But equally, we sought to retain the power to act independently on occasion. The key security relationship would, though, be that with the US. And to achieve influence within US-led coalitions, we would need 'first-day-of-the-war' capabilities. Put another way, we would need conventional war-fighting forces that could be deployed rapidly, in the most challenging of scenarios, and of sufficient scale to be of use to – and thus gain influence with – the superpower. The force balance arrived at in the SDR reflected these judgements. And the force size was largely decided by the decision that Britain wished to have the capacity to conduct one large-scale operation, with some notice, or two medium scale operations but for no longer than six months concurrently. These decisions largely sized and shaped the UK's force structure for the first decade of the third millennium.

But perhaps the key point was that, whilst we had decided on the size, shape and capabilities of the forces that we would want to deploy, we had less idea how we would use these forces to improve our national security. The implicit – and unchallenged – assumption was that the UK would prefer to address security issues by playing our military matches 'away', rather than at 'home'. But the likely tactics for these matches were something of a mystery. There was no mention, for example, in the *Strategic Defence Review* of counter-insurgency.

The MoD reacted to 9/11 with a quick but sophisticated analysis, known as the *SDR New Chapter*. This sought to understand the nature of the new international terrorism, and consider what sort of military responses would be needed. This work led, in turn, to the White Paper, *Delivering Security in a Changing World*.[2] Although there would initially be a slight uplift in defence spending on the heels of 9/11, the policy changes were largely effected through 're-balancing' – i.e. shifting resources – rather than significant additional investment. The 'rebalancing' was (generally) away from 'heavy' conventional war fighting capabilities, towards those that were lighter, more rapidly deployable, but also employable in a wider range of envisaged scenarios. We also

---

[2] UK MoD, *Delivering Security in a Changing World: Future Capabilities.*

sought to develop a better understanding of the expeditionary roles that we might use the forces in abroad, and thus deduce a more sophisticated force structure design. That said, there was in *Delivering Security in a Changing World* still no look-in for counter-insurgency.

A second key theme was the desire to get better leverage from our forces through improved information sharing, using military internets. Known as 'data links' in navies and air forces, the military internets were grouped under the (clumsy) heading *Network Enabled Capability*. This chimed with US thinking under the (slightly more attractive) banner of *Network Centric Warfare*. More importantly, it also chimed with the view amongst radicals in the US defence establishment who believed that a revolution in military affairs (RMA) was underway and leading to a new form of war. In this new form, light land forces, network-enabled, moving at speed, and supported by overwhelming naval and air firepower, would defeat larger but more ponderous forces of enemy states, quickly, and at low relative cost. The first – 100 hours – Gulf War had but pointed the way. Eerily, a key supporter of this new thinking was Donald Rumsfeld. A key statement showed that we had failed to anticipate the likely nature of operations in Afghanistan and Iraq:

> The assumption, based on the experience of the last 10 years, that on enduring operations, once the joint force has been deployed and stability established, lower force levels and generally lighter forces are required.[3]

Implicit in this statement was the idea that stability would be won quickly, soon after joint force deployment. Within months, experience in Afghanistan and Iraq would show that this was badly flawed.

The force structure and capability changes envisaged in the *Delivering Security in a Changing World* would be implemented from 2004, with Afghanistan and Iraq as backdrop. And the British armed forces would continue their route to becoming more joint, lighter, more rapidly deployable, and certainly leaner. But within this thinking, one particular curiosity continued unnoticed. The downsizing of the RAF and the RN had been very significant – personnel numbers in both services had reduced by nearly a 50% during the post-Cold War period. But notwithstanding that our commitment to Northern Ireland – at its height between 25,000 and 30,000 soldiers – had begun to reduce after the Belfast Agreement in 1998, the proportionate reduction in the British

---

[3] Ibid., 3.

Army had been small over the two decades. Why, with a Northern Ireland draw down, and with the loss of a major Soviet foe and the removal of the prospect of conventional war on the Central Front, had Britain retained a large (in British terms) standing Army?

There is room here for no more than speculation, but an entirely plausible answer would be a mixture of stealth and effective institutional lobbying. Unlike the other two services, the modern British Army's domestic footprint was broad. Line Regiments had well-established links with Britain's regions, and the Territorial Army reservist further extended the web of influence. And, again unlike the other two services, a number of middle seniority Army officers had found their way into politics – not a deliberate Army policy, but a useful pathway, nevertheless, for influence. But to justify this large standing Army, it needed to be used. And to be used, it needed a Prime Minister with a penchant for acting abroad, a Prime Minister with an interventionist inclination.

British Post-Cold War Defence Policy – As Practised

British defence planners established the size, shape and capabilities of the UK Armed Forces, but were these forces used in the way our policy envisaged? The test would be events. In simple terms, the defining events of the 1990s were the stabilizing interventions, often in the Balkans, to deal with the ramifications of the end of the Cold War. Whereas the defining events of the 2000s have been the counter-terrorist (and subsequently counter-insurgency) interventions in Afghanistan and Iraq to reduce the threat to the West of terrorist attacks by Islamic extremists.

The early years, as the peace dividend was being realized, saw a large-scale deployment to Iraq, in 1991 for *Operation Granby*, centred on a British armoured division and a proportionately large RAF component. The early policy analysis had not foreseen this contingency,[4] but fortunately, with the Cold War only just over, there was sufficient resilience within the force structure to make this deployment feasible and successful. The first Iraq War was a classic state-on-state war, but testament also to the immense advantage that modern air power brought to conventional land campaigns. After the hammering of the Iraqi forces had

---

[4] Famously on display outside the Chief of the General Staff's office in the MoD in the 1990s, was a policy judgement made just before Iraq's invasion of Kuwait that the MoD could envisage no circumstances in which a British armoured division deployed 'out of area', i.e. outside the NATO operating area.

taken place in the preliminary air campaign, the subsequent hundred hours land campaign if not a walk over, was very nearly so.

This conventional war also bid farewell to such a conventional form of war, and the rest of the 1990s settled into a decade of complex emergencies. These were not state-on-state actions, but rather muddy and difficult instabilities – 'new wars', in Mary Kaldor's view[5] – in which the challenges of post-Cold War peacekeeping were writ large, notably in the Balkans. Although the Former Republic of Yugoslavia was on the opposite side of Europe, Britain – and in particular the British Army – played a prominent and generally successful role in these conflicts. The success was founded in part on the doctrine and capabilities that the Army had developed in the hard training ground of Northern Ireland, and in part on the innovative use of modernized Cold War capabilities. By the time of the SDR in 1998, there was thus a significant data set of generally successful operational experience for the MoD to draw on, and many of the judgements in the SDR reflected this hard won experience. But whether, as a result of these military deployments, the overall sum of British national security was increased is a mute point – and one to which we will return.

The Blair Government brought conviction politics to Britain's foreign affairs. As a House of Commons Research paper suggests:

> The inclination of Prime Minister Tony Blair toward a more multilateral and interventionist foreign policy has had a fundamental impact on both the direction of defence policy and on the role and nature of the British Armed Forces.[6]

One of the key lines in the SDR was that the British Armed Forces should be a 'force for good in the world'. Although this statement was never properly defined, it ought not surprise us that, with this policy intent and a Prime Minister of an interventionist inclination, the British armed forces would be busy. As such, the pursuit of an interventionist defence policy satisfied at least two mutual institutional interests.

The Blair years naturally divide into the pre- and post-9/11 periods, with three operations – in Iraq, Kosovo and Sierra Leone – defining the former, and two – in Afghanistan and Iraq – defining the latter. In 1998, UK forces in a small US-led coalition struck Iraq in October in

---

[5] Mary Kaldor, *New and Old Wars: Organized Violence in a Global Era* (2nd edition, Stanford, CA: Stanford University Press, 2007).

[6] Taylor S. Gick and T. Waldman, *Research Paper 08/57, British Defence Policy since 1997* (House of Commons, London, 2008).

Operation Desert Fox, using air power and sea launched cruise missiles. The explicit political objective – and international justification – was to write down Iraq's ability to build weapons of mass destruction (WMD). The implicit aim was to weaken Saddam Hussein's hold on power. The operation was judged a success after four days, with no significant coalition casualties. In 1999, NATO air forces attacked Serbian targets between March and June in Serbia and Kosovo, in *Operation Allied Force*. The political objective was to force Milosevic to desist ethnically cleansing Kosovo. The impact of the air strikes is a matter of debate but when, late in the campaign, large NATO ground forces prepared to deploy into Kosovo, Milosevic gave way. Finally, in 2000 a UK National Evacuation Operation (NEO) between May and June morphed into an operation to stabilize Sierra Leone. Based initially on a Parachute Battalion, relieved in place by a Commando Battle Group supported by carrier air power, *Operation Palliser* was a conspicuous success. Short, sharp, effective and cheap, the campaign avoided any messy medium or long-term engagement, and fitted perfectly with SDR thinking – 'Go first, go fast, go home.'

What were the common features of these operations? Three stand out. First, they were short. Second, casualties were all but non-existent. Third, because of the first two characteristics, the political capital expended by the Blair Government was tiny. Indeed, by the completion of *Operation Palliser* in mid-2000, there is an argument to say that the combined impact increased Tony Blair's political capital, and Britain's international standing, both in the US and Europe. And the way in which UK forces were deployed and employed in these operations seemed to confirm the SDR analysis. It seemed Britain had got its defence policy right. But would its policy survive the strategic shock of 9/11?

The United States' early successes in Afghanistan, post-9/11, led key American decision-makers to deduce dubious lessons. The Northern Alliance, supported by CIA operatives, US Special Forces and overwhelming Allied (predominantly US) air power, had deposed the Taliban in an unexpectedly short contest between October and December 2001. A hostile government had been toppled with minimal casualties, and without the need to deploy large land forces. This uneven contest thus seemed to support the 'Rumsfeldian' view of new light war – at least in Neo-Cons circles in the US. But there had been, simmering under Afghanistan's political surface, a nascent civil war, with the Pashtun Taliban pitted against the non-Pashtun warlords and tribes of Northern

Alliance. And the course of events had had much more to do with this political context, and much less with the way the war was waged. But this essential lesson was missed, and instead the spurious new light war lessons were more grist to the mill for the Neo-Cons – who had unfinished business in Iraq.

Britain's operational responses to 9/11 were swift. Forces were deployed to support the US operations to root out Osama bin Laden and Al Qaeda (AQ) from the Pashtun areas of Afghanistan. The early Afghanistan campaign had, for the British, been a success. British Special Forces, and then mountain-warfare trained Royal Marines, had deployed and worked successfully with their US colleagues. But as Iraq loomed, the US took their eye off the Afghan ball. And so too did the British. Critically, the US, British and Coalition allies missed two key points. First, the Taliban had been deposed, but not defeated, and they were regrouping. Second, with their Ghilzai Pashtun lineage, it would be in the Pashtun provinces of Southern and Eastern Afghanistan that they would seek to return to the contest. The calm and stability that the British had experienced in Mers el Sherif in the Northern Alliance areas was, thus, not a good indicator of what NATO might expect as it moved south to the Pashtun provinces of Kandahar, Helmand, Uruzgan and Zabul.[7]

Iraq doubled the stakes and it undermined a key British defence policy assumption, namely that the country would not undertake two medium scale deployments concurrently for more than six months. With a medium scale operation underway in Afghanistan, the UK now embarked on a large scale one in Iraq, centred on a British Division in what would become known as Multi-National Divisional area, South East, or MND(SE), in Basra. Most thought that, after the experience of the first Iraq War, overthrowing Iraqi forces would be straightforward. And so it was.

Yet, notwithstanding the new light war views of Donald Rumsfeld and the Neo-Cons, it seemed clear to most thoughtful military people (including many in the Pentagon) that the challenges and manpower requirements for Phase IV – stabilization – of the operation would be significant. And so it proved. More importantly, notwithstanding a positive start to stabilization in 2004, by 2007 the British were in deep

---

[7] I recall visiting Kabul in early 2005, with the UK Chief of Defence Staff, to be told by a British general that Afghanistan was, in terms of security incidents, 'no worse than a bad day in Northern Ireland'.

trouble in Basra. Britain's declared 'strategy' had been to conduct, over time, a transfer of control of the four provinces of MND(SE) to Iraqi security forces. But as any intelligent student of strategy would observe, this was not a strategy for success, but rather one for disengagement. Although cast as 'conditions based' – i.e. that provinces would be handed over when the security conditions permitted – this was not how matters played out in practice. In Basra the handover proceeded, despite the rapidly deteriorating conditions. By 2007, the Jaysh Al Mahdi (JAM) were hounding British forces incessantly in the city and on their operating bases. As Air Chief Marshal Stirrup explained to the Chilcott enquiry in February 2010, the MoD concluded that the British had lost the consent of the people of Basra, and were now mired in a struggle for power amongst different Shia factions.[8]

But as Petraeus and the US surge showed – given political will, military courage, and intelligent doctrine – there were other options. So whilst the British conducted a 'retreat-in-contact', the US took the bull by the horns, surged into Iraq, and snatched victory – or at least breathing space for the Iraqi political process – from the jaws of defeat. Fortunately for Basraris, the Maliki Government, with US support, saved the day. In the *Charge of the Knights* operation, in March 2008, Maliki and the Iraqi Army, with US support, kicked the JAM out of the city. *Charge of the Knights* perhaps turned the campaign, but it did so without the British. By this stage, most in Iraqi Government, many in the US, and the more thoughtful in the UK military believed that the British had been defeated in Basra.

Was it a defeat? This depends on your view of what Britain's political objectives were in Iraq. The popular debate rages over two alternatives. The UK was there to either remove weapons of mass destruction, or affect a regime change. However, if you judge that Britain's overarching objective was to be a good ally to the US, then the answer is clear. As the precipitate British withdrawal from Basra had not been what our US allies had wanted, Basra was a political and military defeat. But at least it allowed the UK, in very senior military eyes, a 'better go' at Afghanistan, in a rural province – Helmand – where the conventional strengths of UK forces, including its air power, could be brought to bear clear of the urban mire. Here too, though, miscalculation seemed to be the order of the day. Intelligence estimates that the province contained

---

[8] The Iraq Enquiry, http://www.iraqinquiry.org.uk/transcripts/oralevidence-bydate/100201.aspx#am (accessed on 15 March 2010).

just 400 or so Taliban proved wide off the mark and, by late 2006, UK forces had become engaged in a counter-insurgency as intense as anything since the Second World War – and from which there is, as of the end of 2010, no sign of let up.

Can we link the operations of these two decades? I think we can. Although the terrorists have had some spectacular successes in 'noughties', at no time in the last twenty years has the survival of Americans or Europeans as sovereign nation states been in question. In part as consequence, the post-Cold War conflicts share two characteristics. First, unlike the three great struggles of the twentieth century, the objectives that British, and the coalitions in which the UK has acted, have sought to deliver have generally been to do with national security rather than national survival. Second, because the objectives have been to do with national security rather than national survival – and thus inevitably more complex to deduce – there has been a considerably higher degree of discretion on whether or not to proceed. These have been wars of choice, rather than wars of necessity.

In the long view, it is the Twin Towers attacks that divide the period. After 9/11, the UK supported the US in a policy of counter-terrorist interventions in Afghanistan and Iraq. These interventions have not played out as expected – the hoped for short wars have turned into long complex counter-insurgencies. And although we have come round to the classic counter-insurgency precept of needing to get boots on the ground, we have found that boots matter little if the higher level politics are not right. Last, but not least, the costs have been higher than expected, not just in blood and treasure, but also in national prestige and political capital. Ultimately, for the British, the pre- and post-9/11 operations are delineated by success and cost. Generally, for all the complexities, difficulties and setbacks, the interventions of the 1990s were both successful and inexpensive. Whereas the interventions of the 2000s have produced mixed results, for which we have paid dearly – and the overall national security benefit for Britain is by no means clear.

The National Security Question

How has our national security changed over these last two decades? In the Green Paper of February 2010, the MoD set the scene for a second Strategic Defence Review. Within it, there is a bold statement:

The success of the (Afghanistan) mission is of critical importance to the security of British citizens and the UK's interests (. . .).[9]

This use of the term 'security' – implicit shorthand for national security – is widespread in official and popular writing, but unhelpful for as long as the term remains ill defined. We need to do better. We need a more tangible sense of what national security is, if we are to reach preliminary judgements on how this has changed over the last twenty years as a consequence of our military actions abroad. Although at its heart, the term security seems to be as much about subjective national perception as it is about objective national reality, we can nevertheless get a better intellectual purchase on the idea by considering what might be termed the 'core components' of national security.[10]

The core components can be identified as follows. First is your physical and geographic security, including that of your territory, national infrastructure and state institutions. Second is your political security, including that of your political systems and processes, and the maintenance of national values, such as freedom of speech. Third is your economic security, in particular your ability to create and maintain national wealth, and to trade. Within each component area, the security will be a function of both the threats and your ability to defend against them, both at home and away. This simple framework is not a panacea. But it should nevertheless allow us to apply a touch more analytic rigour as we consider the degree to which our security has changed – relatively – over the last twenty years as a result of our interventions. And it should thus help us get a better sense of the strategic value of these interventions.

Are the streets of London – or Washington, Paris, and Berlin for that matter – safer? Has the security of our politics improved or worsened? Are our values better or worse protected? Has our wealth been affected positively or negatively? Is our ability to trade internationally more or less secure? With space for no more than the most cursory examinations, let us focus on the relative change in national security that might be attributed to the Afghanistan and Iraq interventions, both justified in its name? How does the credit and debit side of the national security ledger look as a result of these first wars of the third millennium?

---

[9] UK MoD, *Green Paper: Adaptability & Partnership: Issues for the Strategic Defence Review* (London: MoD 2010), CM 7794, 13, http://www.mod.uk/NR/rdonlyres/790C77EC-550B-4AE8-B227-14DA412FC9BA/0/defence_green_paper_cm7794.pdf (accessed on 15 March 2010).

[10] These ideas apply, albeit at smaller scale, to Britain's sovereign overseas territories.

On the credit side, there is a modicum of modern government in Iraq. Operations in Afghanistan – and latterly Pakistan – have had a disruptive impact on AQ's bases and training. There have been some terrorist successes post 9/11 – the Madrid rail bombings on 11 March 2004 and the London transport bombings on 7 July 2005 – and some near misses. But since then AQ and its affiliates have failed decisively to penetrate the West's home defences. Whether this is because of improved physical security on the Western mainland and entry points or the disruptive impact of our military operations abroad is, though, a mute point – but military instinct leads me to conclude that home defence is the more dominant factor.

The West have relearned how to do counter-insurgency, and the US has prevailed in the face of AQ and militant Islamic forces in Iraq. Indeed, in Western militaries, experience of counter-insurgency operations has never been higher and if, once the Afghanistan campaign is concluded, we wish to continue with an interventionist defence policy, then we have the military forces and experience so to do. But perhaps the most important entry on the credit side is one of insight – the West now has a much better understanding of the potential but also the limits of its power, and of the complications of interventions in Islamic countries. As a result, the Neo-Con experiment has probably run its course.

On the debit side, tactically the West's sophisticated armies' vulnerability to unsophisticated insurgent tactics have been exposed, with the IED their Achilles' heel. Strategically, Western forces have been fixed in Afghanistan and Iraq, and their ability to act elsewhere in the world prejudiced. And there has been a corresponding reduction – very significant amongst UK forces – in routine training for roles other than counter-insurgency. And as a result, the capacity of our forces to protect us by acting in these other roles has significantly degraded. But perhaps the most uncertain and biggest debit entry is the extent to which Western military actions in Muslim lands – whether or not it is with the concurrence of the host governments – and the attendant press coverage have helped extremist Islam's 'recruiting sergeants', and thus increased the overall potential threat. Over a fifth of the World's population follow Islam, and the gearing from even a tiny increase in percentage recruitment would thus be significant. If just an additional 1% of young Muslim men were so influenced, then ranks of potential Jihadists would be swelled by over a million.

On balance, the West's national security account may be slightly better off, at least in the short term. This is because an unintended consequence of these two operations seems to have been to 'fix' extremist Islamic attention in Afghanistan and Iraq, allowing the West to significantly improve its homeland defence. But it is a finely balanced judgement, not least because of the 'recruiting sergeant' debit. If AQ's success in radicalising young Muslims has been significantly aided by the West's expeditionary operations, then notwithstanding the short-term credit, in the medium to long term the debit may be significant.

How much has it cost us to secure this short-term credit, if credit it is? Our treasuries have paid in treasure. Stiglitz and Bilmes calculated in 2008 that the total costs for UK of the Afghan and Iraq Wars by 2010 would be of the order of £20 billion. And that, for the US, when the two wars are taken together, only the Second World War has exceeded their combined financial total.[11] And our militaries have paid in blood. Over 500 British servicemen have been killed in action in Afghanistan and Iraq, but this figure is the tip of the iceberg when compared to those who have suffered life-changing injuries, and are no longer available for active service. The political costs have been equally high. Many would argue that, after Iraq, the reputation of at least one UK Prime Minister lies in tatters. And both conflicts have been festering sores in the political hide of administrations in the US, UK, Canada, Holland and Germany and led to the collapse of the Dutch Government in 2010. Which, in turn, has all but destroyed the popular appetite for expeditionary operations such as these in our short- to medium-term future.

Nevertheless, the first fundamental question – did the interventionist policy work? – remains unresolved. But a second is of more immediate interest. Do we want to continue to structure our defence policy around the idea of sovereign intervention, stabilization, and counterinsurgency? If our guard is now up at 'home', do we want or need continue this way 'away'? If so, after Afghanistan, where next? If not, is there another way?

The Current Strategic Context

The current strategic context shaping future defence policy is one in which the short term is largely dominated by Afghanistan and the wider threat of Islamic terrorism. Yet, an attentive observer can also

---

[11] J. Stiglitz and L. Bilmes, 'Three Trillion Dollar War', *The Times*, 23 February 2008.

now see beyond, into the medium and long term. Notwithstanding the pressing existential threats of the short term, history suggests that the UK would surely be foolish to prepare solely for these 'last wars' and counter-insurgency. Earlier on, the Green Paper of February 2010 was quoted, with a promise to return to it, in particular to unpack one important assumption on Afghanistan:

> Our Armed Forces are in Afghanistan to protect the UK's national security by denying safe haven to violent extremists. *The success of the mission is of critical importance to the security of British citizens* and the UK's interests, including the reputation of our Armed Forces. We will have succeeded when the Afghans themselves are able to prevent and suppress violent extremism within their borders. [emphasis added][12]

If the analysis of the national security ledger presented in this chapter is anything to go by, then the emboldened assumption is questionable. The assumption also fails on two other tests, both reasonably straight-forward. First, if Afghanistan is critical to British national security then, should the US decide to draw down in 2012 before our criteria for success have been achieved, presumably we would stay on? Second, if the logic of the intervention is to deny safe havens to violent extremists, then after Afghanistan we will no doubt be moving to one or more of the other safe havens, such as Somalia, Yemen, or Sudan? There is, of course, no such talk, but this is hardly surprising given the lack of polit-ical and popular appetite for any such action. So, on the logic of these three tests, the success of the Afghan mission is important to British security, but not critical.

It is important for three reasons, none of which are directly con-nected with denying safe havens to violent extremists. First, for as long as the US are engaged in Afghanistan, then Britain's engagement is an opportunity to repair the 'reputational' damage of Basra, and sustain the special relationship, with all its security benefits. Second, for as long as NATO is engaged in Afghanistan, then Britain's engagement helps discharge our responsibility to the most important of our international security alliances. Third, and more uniquely for Britain, our engagement in Afghanistan in the Taliban's Pashtun heartlands north of the border is complementary to Pakistan's engagement in those to the south. For these reasons, Britain needs to stay engaged in Afghanistan. But we need to stay engaged on the understanding that the campaign is not critical to British security and does not, therefore, require us to sacrifice

---

[12] UK MoD, *Green Paper: Adaptability & Partnership*, 13.

every other element of defence in its pursuit. Rather, British investment in the campaign ought to be proportionate, based on these imperatives.

The second consideration to help us decide on the scale of our future Afghanistan investment is that the British armed forces are, in relative terms, bit players. The USMC already outnumber us in the Helmand province. Afghan wide, the two majority actors are, by some distance, the Afghans and the US. Success or otherwise will thus depend on their actions. So, although there is much talk of more British boots being needed on the ground, the mathematics tell the unquestionable strategic story.[13] An additional 1,000 or so British soldiers, and the expense incurred, placed in one half of one of Afghanistan's thirty-four provinces, will make less than 0.5% proportionate difference to the overall force size and is thus unlikely to make a significant strategic difference to the campaign.

The third and final consideration to help us decide on our Afghanistan investment is that, ultimately, the way that the West takes the campaign forward will be decided in Washington. It is perhaps just as well, therefore, that it is within the Beltway that the most intelligent recent discussion on campaign strategy has occurred. For the moment, the counter-insurgency boots-on-the-ground advocates, such as General Petraeus, have won out over the counter-terrorist 'lite' advocates such as Vice President Biden. Although the President has agreed to an Afghan surge, he has done so within stipulated time constraints. This is not remotely surprising, given that Washington's view will be critically influenced by the way the campaign plays out in US domestic politics. Which US President – with health reform as a key aspiration and the US economy as a key challenge – would want to enter a re-election year in 2012 with a Vietnam-like albatross hung round the neck of his candidature. The clear deduction is that the timelines of the campaign are largely outside Britain's control, and we must plan our investment with this consideration in mind.

In sum, it will be right to see Afghanistan through for as long as the US and NATO are engaged. But ultimately our investment should reflect the unavoidable conclusion that the campaign is not critical to Britain's national security – be it territorial, physical, economic, or political – and thus be sized and shaped accordingly. In shaping our commitment, we should seek to engage those forces that are of most

---

[13] See, for example, the Chief of the General Staff's views in M. Smith, 'Top General Calls for New Cyber-Army', *Sunday Times*, 17 January 2010.

help to US and Afghan allies – specialist and niche capabilities, such as special forces and well found training teams, offer the most strategic leverage. And in sizing our commitment, we should select a level that we can comfortably sustain in the field without bleeding the remainder of defence dry, and thus critically weakening our national security in other areas.

The broader threat of militant Islam is never far from contemporary Western security consciousness, and there is no shortage of failed or failing states where AQ and other extremist Islamists have toeholds. This is particularly the case in and around the Horn of Africa – Umar Farouk Abdulmutallab, the young Nigerian man who attempted to detonate a bomb aboard an Amsterdam-Detroit flight over Canadian airspace in December 2009, was schooled in Yemini-training camps. But before our counter-terrorist and counter-insurgency reflexes lead our thinking in these directions, we would be better observe the matter at a higher, grand strategic, level.

The most important thing, before the UK decides on measures to take, is to ensure that we really understand the political context in which Islamic extremism exists, and that we get to the bottom of the political issue at contest. There is space here for only the most cursory of examinations, but we might best describe the context by saying that, as Islam collides with modernity, the West finds itself as an unwitting party to an Islamic civil war of ideas, between modernizers and conservatives. The modernizers believe that the problem is that Islamic nations have failed to keep pace with the West's modernization. Whereas the conservatives believe the problem to be too much, rather than too little, Western modernization. The West has been (largely unwittingly) drawn in on the side of the modernizers, and has hence become a target of Islamic terrorists, who are the conservative extremists.[14] So because politics and Islam are indivisible in the critical Muslim nations, the political issue at contest is the nature of governance in the Islamic nations of North Africa and the Middle East, an issue that will take time – decades as a minimum – to play out. And, whatever our chosen course of action, we need to tread with great cultural care and sensitivity.

What is the severity of the threat that we in the West face, when compared to, say, that which existed before 9/11 or in the years immediately

---

[14] I am indebted in these thoughts to discussions with my brother, Lt Cdr R. A. Jermy, RN, an Arabist, Arab linguist, and veteran of both the Iraq and Afghanistan campaigns.

following? As we ponder this question, it is worth recalling that a general lesson in history is that terrorists are very rarely defeated militarily. The more normal experience is that, through a mixture of offensive and defensive measures, the threat is reduced to a level that is both acceptable to the general population, and unable to interfere with the politics at stake. It is always a difficult judgement to decide whether this state has been reached, but the lack of decisive AQ attacks in the past five or so years, suggests that we could be close to this level. If so, then now might be the moment to search for a sustainable policy that we can settle into for what will be an generational marathon, stretching far beyond Afghanistan's conclusion.

Furthermore, as we look into the medium and long term, it seems clear that strategic challenges of the future will not be limited to Afghanistan and Islam's civil war. This is not the place to speculate on the full range of possible challenges, but four other factors seem likely to shape the strategic future. The first will be the policy of the United States and of China (the only other credible contender for superpower status in the medium term). Second, as long as oil is the world's primary energy source, the stability of the Arabian Gulf, and other oil producing regions, will remain an area of legitimate national interest for those who consume oil. Third, although their internal contradictions will deny them superpower status, Russia and India are the heavy weights of their regions. Fourth, and most importantly, as the inevitable consequences of global warming play out in the coming decades, the chances of political conflict as a result of resource competition, migration, or the search for habitable land seem likely to rise. The list could be much longer. But for factors which might bear on future armed conflict or wars – particularly significant international wars – the author would give short odds on one or more of these four being implicated. And if so, then our defence policy needs to be shaped with these considerations in mind, particularly the last one – global warming – around which there is most uncertainty.

Consider the picture painted by Michael Lynas on the impact of a global rise in temperature of 3°C, well within the sensitivity of current global climate modelling scenarios:

> With structural famine gripping much of the subtropics, hundreds of millions of people will have only one choice left other than death for themselves and their families: they will have to pack up their belongings and leave. The resulting population transfers could dwarf those that have historically taken place due to wars or crop failures. Never before has the

> human population had to leave an entire latitudinal belt across the width of the globe.[15]

No one can yet allocate a probability to this scenario, but it is entirely plausible and by no means the most apocalyptic of the global warming possibilities. And as such, it perhaps puts the threat of Islamic terrorism into perspective. This may be the time to think about appropriate polices and strategies, and not put all our eggs into the tactics of the counter-insurgency basket.

An Offshore Defence Policy

One obvious alternative to an interventionist defence policy is what one might, for the sake of argument, entitle an 'offshore' defence policy. The fundamental principles of such an approach would have been well understood by Sir Julian Corbett, because many would be drawn from Britain's successful maritime defence policies of the nineteenth and early twentieth centuries.

Their central ideal was an adaptation of Clausewitz's thinking, articulated by Corbett. In unlimited wars, Corbett agreed with Clausewitz that defeat of the enemy's armed forces would be your primary military object. Whereas in limited wars it might not be. Corbett went on to suggest that in wars fought between continental neighbours the distinction between unlimited and limited war was difficult to sustain. The shared boundary meant that a limited war could easily grow into an unlimited one. Whereas, if the states are separated – by the sea – then the state commanding the sea, could take as much or as little of the war as it chose, as was the case, for example, with the United States in Vietnam.

By adopting defence policies based on this fundamental principle, supported by local forces in the Empire's colonies underpinned with the essential safety net of the Royal Navy, Britain grew and then maintained her Empire on a military shoe-string. Things would change in the twentieth century, as peer competitors grew on the European continent. The need to raise large land forces for the First World War and the Second World War, and then maintain them throughout the Cold War was surely correct. With these existential threats, Britain

---

[15] Mark Lynas, *Six Degrees: Our Future on a Hotter Planet* (London: Fourth Estate, 2007), 159.

needed to maintain a credible and larger standing Army, to contribute to the coalition defence of our interests in our European near abroad. But absent the twentieth century threats of the USSR and the IRA, might there not be merit in an island nation such as Britain adopting, in the twenty-first century, something along the successful maritime lines of the past? And if so, what would such a policy look like, in ends, ways, and means?

Fundamental to the idea of limited war is that of limited objectives – limited wars are those fought for limited objectives. And the fundamental feature of Britain's maritime policies for almost two centuries was that the political objectives sought were generally limited. And so it would be for a future twenty-first century offshore policy. It is not possible to intervene with mass or to state-build from offshore. An offshore policy could not deliver the objective, sought in Afghanistan, of denying safe havens by reducing ungoverned Islamic space through nation building and counter-insurgency. If this is something that we decide to continue to do post-Afghanistan, then there is no need here to rehearse further what is implied. We know well, from Afghanistan and Iraq, the investment of blood, treasure, political capital and time required – and the prospects of success. Rather, an offshore policy would pursue two generic objectives, both more limited. First, it would seek to develop indigenous forces of the host nation – security sector reform (SSR) in the jargon – in order that they were better able to confront and defeat extremists operating illegally within their territorial boundaries. Second, it would seek to contain and disrupt terrorists judged to be a clear and present danger to the West. Although each individual case would need to be judged on its merits and special circumstances, the generic strategy would be two pronged, based on the two objectives.

First, we would develop indigenous forces drawing on the training and security sector reform expertise built up in Afghanistan and Iraq. Because of the gearing, training would become a mainstream military role rather than, as now, something found from within the hide of the fighting force structure. In deciding where to assist Islamic governments, we would be demand, rather than supply, led responding to the requests of host governments, rather than pushing ourselves upon them. But a key factor in our decision to assist would be politics. If there were a working and evolving political process, with adequate consent, and positive human rights intent, then we would help. But we would avoid engagements were we could be cast as propping up unpopular or illegitimate governments, thus playing into the hands of the extremists. When trainers were deployed, then subject to the assessed threat we would support the deployed teams from the air or from the sea or both, with, for example, quick reaction strike forces, fire support, and combat search-and-rescue capabilities.

Second, were we to gain intelligence of terrorist threats that the host government were unable to confront, then we would reserve the right to act against them – but noting that, sometimes, doing nothing is the best strategic approach in these complex political campaigns. And in the event that there was reason to strike terrorists targets, then again our modern experience – using, for example, special forces, amphibious forces, and maritime or land based air power – would be similarly relevant. But if strikes were undertaken, then a key factor would be our footprint, in space and time, with the strong preference being to minimize both.

Two core design principles would underpin this approach. First, the UK would seek to minimize, rather than maximize, Western boots on the ground. This would not be so as to reduce casualties – albeit this would be a likely by-product – but rather to reduce the medium and long-term political downsides of having large bodies of Western troops deployed in Muslim lands. Second, and as corollary, British policy would seek to hold offshore all that could be taken offshore, for example, command and control, air power, fire support, contingency reserves, and logistics. If the metaphor for the counter-terrorist interventions in Afghanistan and Iraq is invasive open-heart surgery, then that for an offshore policy would be twin tracked, nutritional supplements to improve the immune system, in concert with occasional keyhole surgery to root out the cancers.

The key advantage of operating offshore is that the sea provides a rear and forward area that, first, is all but invulnerable when compared to the land and, second, is politically independent. The asymmetric tricks of the terrorist's trade do not work at sea. Indeed quite the reverse. The strategic advantages are thus in force protection in all its senses. Take, for example, the protection of air bases and logistics re-supply. The numbers killed in defence of British air bases in Afghanistan and Iraq are well into double figures. Whereas the last time that anyone was killed in action in a British aircraft carrier was the SecondWorld War. And the expenditure, costs and losses through action against Western main supply routes into Afghanistan and Iraq are in stark contrast to the losses in our sea lines of communication, of which there have been none.

But it is at the political level that the advantages of an offshore policy are manifest. Because territorial waters extend only as far twelve nautical miles, then forces stationed further offshore are both politically independent and out of popular sight. They can be stationed and moved without reference to host nations, poised over the horizon, but available at very short notice if needs be. A more subtle advantage is that you do not need to press a reluctant ally into providing access, basing and over-flight (ABO) for your deployed forces. This is a particular advantage where the state from which you might seek ABO is a friendly Islamic state, as it obviates the need to base the boots of the Crusaders on Muslim lands, and thus removes a major political stick that extremists use to beat indigenous governments.

And finally there is the matter of cost. Clear of the need for costs unique to land basing – such as life support, civil engineering, local contractor fees, corruption and losses to enemy action – running costs would inevitably reduce. Is an offshore policy feasible? Perhaps the capping advantages of such a policy is that for Britain – and America for that matter – the forces and core capabilities needed to prosecute it are already in place, or ordered. First, for the lead tactical roles of training and strike, it is difficult to recall a time when we had more experience. Second, the strategic move offshore would use capabilities that are either already in the inventory, or soon to arrive. Simply put, one needs amphibious shipping and marines, aircraft carriers and air groups, and logistics shipping, together with a modicum of force protection based on destroyers and frigate forces, of which Europe and America are well endowed. So, as well as being politically and strategically attractive, an offshore policy would be militarily feasible and resource 'lite'.

Conclusions

Whilst Afghanistan is up front and personal in current defence policy thinking, one way or another it will come to an end. And when it does, it will likely be to a tempo based on US political cycles – and beyond the control of UK and the major NATO players. The eventual result should thus become clearer over the coming two years. If American military commanders can stabilize the security situation, and President Karzai and the Afghan Government find a way of ruling Afghanistan with less corruption and more consent from the people, then there will be cause to see it through to a successful end. But if not, then the time will come to decide whether we should invest good money on top of bad, or cut the West's losses. And the centre of decision-making gravity will be with the majority shareholder in Washington, and not the minority shareholders in London, Paris, Bonn, Ottawa, the Hague and Brussels.

But whichever of these two futures plays out, one thing seems clear. There will, on the experience of Afghanistan and Iraq, be little immediate popular appetite – and thus political appetite – for subsequent interventions of their scale, intensity, duration and cost. As such, then it may be time to try another way. An offshore policy, focused on security sector reform in failing states (rather than nation building) and disrupting terrorist bases and training is the obvious alternative. When compared to our current interventionist policies, its objectives would be more limited – security sector reform, containment and disruption (rather than defeat) of terrorists, defence of trade, and so on – but then so too would the costs in blood, treasure, political capital, and strategic time.

We have tried open-heart surgery. Now, perhaps, it is time for something less invasive – and less expensive. Now may be the time to move from a *war on terror*, to the *containment of terrorists*, drawing down Crusader boots in Muslim lands, and letting the politics of these countries evolve in ways that are in the grain of their nation's natures.

And such a policy might also be the one best fitted for an uncertain future, where the grand strategic issue of climate change should cause us to re-consider the unwittingly complacent assumption of our National Security Strategy, echoed in the recent Green Paper, that 'there is no direct external threat to the territorial integrity of the UK'.[16] Ultimately, the short-term deciding factor may be cost. Those who see our recent interventionist policy in a positive light cannot deny it has been expensive, even in an age of plenty. In an age of austerity, an offshore policy may not only be the most sensible politically and strategically, but it may also be the most affordable financially.

---

[16] UK MoD, *Green Paper: Adaptability & Partnership*, 13.

CHAPTER NINE

FROM ALLIANCE TO COALITION, THEN WHERE?
JAPAN AND THE US NAVY COOPERATIVE STRATEGY
FOR THE TWENTY-FIRST CENTURY

Yōji Kōda

In exploring the implications of the contemporary security landscape for the defence policy of a maritime country like Japan, the
first thing that comes to my mind, as a former senior naval planner, is the USN document *A Cooperative Strategy for 21st Century
Sea-power*.[1] This was published in October 2007 with the intent of
clearly setting the context for problems common to all maritime
nations today, and to outline basic concepts that can offer guidance
to approach them. This strategy document is particularly relevant
to the theme explored in the last part of this volume as it addressed
contemporary security issues against the challenges of a growing
'interdependent world'. It vividly describes the typical characteristics
of today's globally connected world, where the degree of interdependence among nations has become closer than ever. As Commodore
Jermy examined in the previous chapter, for the UK the interdependent nature of the present world has entered the very grammar of
security planning, with debates that connect military operations in
Afghanistan to national security through issues such as terrorism
and security sector reform.

What about Japan? This chapter aims to investigate the relationship between maritime power and Japanese contemporary security in
reference to the US Maritime Strategy. It does so by drawing upon the
direct experience of this author as a recently retired fleet commander. The analysis in this chapter is divided in four sections. First, it will
seek to draw together ideas about maritime strategy explored in the
earlier chapters of this book . It will look at the connection between

---

[1] J.Conway,G.Roughhead,T.Allen,*CooperativeStrategyfor21stCenturySeapower*,
17October2007,http://www.navy.mil/maritime/MaritimeStrategy.pdf(accessed on
3 September 2009).

the RN and the IJN first, and between the latter and the JMSDF after the Second World War. Thereafter, the chapter will focus on the content of the US Maritime Strategy and its relevance for maritime nations, including an outline of the current different forms of international cooperation. In the third section, the chapter explains Japan's position regarding alliances and coalitions, and the different roles power projection and expeditionary capabilities play in shaping military contributions to both of them. Finally, the chapter will propose the notions of 'coalition over alliance', and 'coalition over kingdom', as recommendations for future policy actions.

Influence and Traditions: The Royal Navy, the Imperial Japanese Navy and the Japan Maritime Self-Defence Force

Before entering the main subject, it is worth reviewing two core aspects of the relationship between the Japanese and the British naval traditions. This review should not represent a surprise to scholars of Anglo-Japanese relations. As Professor Tohmatsu underlined in his chapter, at the turn of the twentieth century the IJN looked up to the RN as a mentor. Yet, what perhaps many do not realize is that the RN still preserves a rather strong capability of influence over various navies in the world today, including the JMSDF, where the present author spent over thirty-seven years as an officer.

In order to contextualize this influence it is necessary to briefly explain the process that led to the demise of the IJN and the genesis and development of the JMSDF within Japan's post-war security policy. After the end of Second World War, Japan entered a treaty with the United States which has come to be a cornerstone of its national security policy and defence strategy. This security partnership has lasted for almost sixty years since the signing of the original treaty in 1951.[2] The current alliance mechanism, which is based on the 'revised' Japan-US Mutual Security Treaty, concluded in 1960, marked its 50th year anniversary in 2010.[3]

---

[2] On the Japan-US Alliance and its wider political context, cf. Michael Schaller, *Altered States: The United States and Japan since the Occupation* (Oxford: Oxford University Press, 1997); John Swenson-Wright, *Unequal Allies? United States Security and Alliance Politics toward Japan, 1946-1960* (Stanford, CA: Stanford University Press, 2004); Makoto Iokibe (ed.), *The Diplomatic History of Postwar Japan* (Abingdon, OX: Routledge, 2010).

[3] For an analysis of the five decades of security alliance, see George R. Packard, 'The United States-Japan Security Treaty at 50: Still a Grand Bargain?', *Foreign Affairs*, Vol. 89, 2010:2, 92-103.

The alliance can be divided in two main periods. The first lasted for the initial three decades, when the bilateral alliance acted as a key security mechanism for Japan and the Western Pacific (West-Pac) region. As the Cold War was a global confrontation, in a wider sense, along with NATO in the western theatre of the Eurasian Continent, the Japan-US alliance played a similarly core role in the West-Pac theatre, deterring and containing the eastern bloc powers built around the Soviet Union. The second phase of the alliance includes the two decades of the post-Cold War era. During this period, the partnership has lost its global function to contain Communism, but it retained a key security role in the Asian region. In this part of the world, various factors of instability remain, notably the security of the Korean peninsula.[4] Indeed, in addition to this fundamental regional-stabilizer function, the Japan-US Alliance has been re-evaluated as an important means to ensure the security of the widely spread area of strategic importance connecting Southeast Asia to the East Coast of Africa, via the Middle East. In these areas, issues like maritime disputes in the East and South China Seas, piracy and the power shift prompted by the emergence of China are of central significance.

Under these security circumstances, the JMSDF, since its establishment, has planned and built its entire operational force in relation to the scope and functions of the Japan-US alliance. In particular, the maintenance of a close and good working relationship with the USN has always been given the highest priority. Since its establishment in 1952, the JMSDF has received a tremendous amount of assistance and support from the USN, both materially and psychologically. From its modest beginnings in the early 1950s to the mid-1980s, the operational potential of the JMSDF remained in a sort of 'infant stage', and the 'capability gap' between the service and the USN was so large that it was difficult for the JMSDF to catch-up quickly with the USN tactically and operationally. In other words, this gap meant that in the naval sphere, the Japan-US alliance resembled the relationship between a strong father and a small child.

Across the first three decades, The JMSDF concentrated all possible efforts towards the enhancement of the force's capabilities. This enabled it to seize important results by the mid-1980s, when Japan could be regarded to possess, for the first time since the end of the war in the Pacific, a 'balanced fleet' to meet its defence requirements. The new capabilities of the JMSDF covered a wide spectrum of functions.

---

[4] Japan Ministry of Defence (JMoD), *Defence of Japan 2009* (Tokyo: Erklaren, 2009), 225.

The service had now the potential to focus on different types of operations, the know-how to pursue the design and building of new ships, to look after questions of logistics, training, education, as well as research and development. As a result of these all-out measures to build up its capabilities, throughout the 1980s, the JMSDF became an operationally-capable naval force, supporting the partnership with the USN in substantial terms.

One crucial factor underpinned the JMSDF's ability to mature as a professional naval force. When the organization was established in 1952, its senior leadership had set a clear goal in mind, making this new force a navy – not an enlarged coastguard.[5] The first generations of senior leaders of the JMSDF, who, of course, included many former IJN officers, had the determined intention to build the JMSDF as a new navy of Japan. This aim notwithstanding, strong negative Japanese public perceptions from the Second World War coupled with the influence of the new 'pacifist' Constitution of Japan, made it impossible to name this new maritime force as a 'navy'. This was the reality of Japan, only seven years after the end of the war. For that reason, senior leaders of the early JMSDF naturally pursued a basic policy to keep the best possible continuity with the IJN, retaining key aspects of its culture and traditions in training and education. In this manner, the newly born JMSDF had in the IJN a primary organizational model.[6]

In so doing, the JMSDF maintained a strong awareness of the original organizational ties with Britain. The IJN had received a strong influence from the RN for many years. On its way of development, since its foundation in 1867, the IJN sought to merge the Samurai/ Bushido spirit of Japan, with the professional traditions and seamanship skills of the RN. This contributed to the rapid professional growth of the navy into one of the leading maritime powers of the world in the early-twentieth century. Over the next twenty to twenty-five years, Japan and its navy maintained good relationships with other leading nations, as well as other navies, until the mid-1930s. Its cosmopolitan outlook and international profile had drawn significantly from the

---

[5] The history of the role of the USN in the development of the JMSDF is described in James E. Auer, *The Post-war Rearmament of Japanese Maritime Forces, 1945-71* (New York, Washington, London: Preager, 1973); and Naoyuki Agawa, *Umi no Yūjō* (Friendship at Sea, Tokyo: Chūōkōronsha, 2001).

[6] Alessio Patalano, *Kaiji: Imperial Traditions and Japan's Post-war Naval Power* (unpublished PhD thesis, King's College London, 2009).

British experience. After that, unfortunately, and in spite of some daring operations fought by the IJN in the Pacific, the navy disappeared from history, through total defeat, after three years and eight months of consuming war. In its seventy-five years of history, the IJN had grown from a small coastal force to one of the world's most impressive and combat-hardened forces, to fade away with nothing left except good traditions and culture. These, in some sense, constituted a positive psychological legacy that traced its roots in the RN.

British naval traditions as they were received and transplanted from the RN into the IJN became a central part of the imperial naval heritage and, in developing the JMSDF, post-war naval leaders did not reject them. Rather, they transmitted them to younger generations of JMSDF personnel. The founders of JMSDF really believed so, and did so for years. Today, all sailors of JMSDF are deeply proud of this legacy that connects them to the two legendary navies, and have strong confidence upon them. It was this invaluable cultural heritage that helped the JMSDF to make the best of the assistance and support from the USN. Very strict practical training was given to the JMSDF units by combat experienced USN training teams from the very beginning. However, even during the then still very infant days of the new Japanese naval service, USN instructors had no need to say anything on the naval fundamentals: sense of duty, war-fighting spirit, seamanship, tactical sense and training concepts. In the JMSDF, fifty years on, officers like myself as well as each sailor in the fleet, surely understand the fact that the origins of our good navy heritage were the indirect products and treasures of the RN – and I for one am very proud of it. This is the essence of the relationship between the JMSDF and the RN that I wish to emphasize in the context of this volume.

A Cooperative Strategy for 21st Century Seapower: A Brief Analysis

In its introduction, this strategy recognizes the fact that security and prosperity in the contemporary world are a product of increasing interdependence among nations. It equally points out that this very interdependence makes national systems vulnerable to disruptions not directly connected to them. The document further mentions the importance of the maritime domain, as well as of maritime power, as vital components to the sustainability of interdependence as to its security. As a result of the interdependent nature of security issues, the conclusion of this chapter is to emphasize the importance of multinational cooperation to establish common approaches to prevent disruptions and guarantee prosperity.

In relation to these preliminary considerations, the document goes on to present the challenges to international security, pointing out how top security issues have an increasing transnational nature. Competition for natural resources, trans-border criminal activities, proliferation of Weapons of Mass Destruction (WMD), natural disasters, are all fast becoming potential sources of conflict. Thus, '(n)o nation has the resources required to provide security throughout the entire maritime domain. Increasingly, governments, nongovernmental organizations, international organizations, and the private sector will form partnerships of common interest to counter these emerging threats.'[7]

The strategic concept unfolding from the above is one openly seeking to take advantage of two forms of integrated maritime operations: alliance-based operations, and informal arrangement-based initiatives, typically represented by the concept of the '1,000-ship navy', re-defined in the document as 'Global Maritime Partnership'. These forms of cooperation are to contribute to the USN global effort to protect the maritime commons as a way to achieve its six strategic imperatives: limiting regional conflict; deterring major power war; winning the United States wars; homeland defence; better relationship with more international partners; and containing local disruptions.

In terms of implementation of the strategic concept, the USN's six core capabilities, comprising its forward presence, deterrence, sea control, power projection, maritime security and HADR (Humanitarian Assistance and Disaster Response) are key enablers to national security and to the build-up of partnerships. Indeed, this Cooperative Strategy – as the name suggests – is geared towards the importance of cooperation and mutual support among diverse elements of the greater maritime community, within (this is the first US maritime doctrine penned jointly by the USN, the US Coast Guard and the US marine Corps) and outside of country. It is in this vein that the document concludes that American seapower 'is a force for good, protecting this Nation's vital interests even as it joins with others to promote security and prosperity across the globe'.[8]

---

[7] Conway, Roughhead, Allen, *Cooperative Strategy for 21st Century Seapower*, 7.
[8] Conway, Roughhead, Allen, *Cooperative Strategy for 21st Century Seapower*, 18.

This new strategy has one prominent characteristic that casts it in a different context from prior USN strategic documents. That is the call for both transnational cooperation and interagency coordination as key enablers to gain a full and thorough understanding of the maritime domain. In this respect, it is not too much to say that the basic concepts of this US Cooperative Strategy, especially this characteristic, are not limited to the United States alone. They constitute a universal truth for all nations. In particular, as this strategy makes clear, 'coalitions' are set to become the most common form of international cooperation at sea.

From the viewpoint of a former Japanese strategic planner, a coalition is a relatively new concept which will test the limits and adaptability of the alliance with Washington under today's complex international situation. An advantage of a coalition over an alliance is that, depending on the incident at hand, one nation is able to count its national interest first, and then decide about its participation in an international effort put in place to settle the incident (the coalition), adapting its involvement to its domestic circumstances. A coalition provides a good amount of political flexibility for participating nations. On the other hand, an alliance, in general, is tailored for a specific purpose. The alliance is firm and good as it establishes 'common' and 'hard' objectives to protect, but less flexible to react to unpredictable and rapidly changing situations. In order to quickly respond to the dynamically changing situations by bringing all possible nations together, an alliance sometimes exposes its limitations, and other forms of cooperation – typically represented by a coalition – are regarded to have some advantages over an alliance today; not fully but just to make up for the limitations mentioned above.

In all, an alliance is one form of international cooperation that has firm objectives to establish common and mutually advantageous forms of cooperation, under strictly binding responsibilities and burdens. By contrast, a coalition is a form of international cooperation that appreciates a large degree of independent judgement from each participating nation, under less binding responsibilities and burdens. So, for most of the nations which have some intent to participate in international cooperation today, a coalition is more desirable and attractive than an alliance from the view point of the political and diplomatic freedoms of each nation.

Any assessment of each form of cooperation, however, should be complemented by an historical account contextualizing their relative advantages and limitations. In this sense, during the Cold War, the nature of the East-West struggle was such that it favoured the formation of 'alliances'. Structurally, the systemic confrontation was based on a polarized disagreement over ideological, political and economic postures between two different blocs. In order to fully and properly respond to this extremely delicate and confrontational security environment, the two blocs formed alliances. In the Western bloc, NATO and the Japan-US Alliance brought together those who, in different regions of the world, stood at one end of the extreme. In the Eastern bloc, the Warsaw Pact stood at the opposite end of this ideational, political and military confrontational spectrum. The objective of each alliance was simple and clear. First and foremost, they were designed to deter 'full-scale' head-on military collision. If deterrence failed and the confrontation escalated into real war, then the alliances were to offer the necessary military infrastructure to gain victory over the opponent. All participating nations of the alliances bore clear, but heavily binding burdens and responsibilities, to pursue the common objectives.

The international situation in the first decade and a half after the end of the Cold War, on the other hand, has shown some important differences with the past, with cultural, religious and ethnic issues replacing ideology as core reasons for international conflicts. Violent conflict among and within small developing nations, illegal trans-border crimes as well as acts of violence carried out by non-state groups replaced the deterrence-based operations conducted by large, opposing military formations. With the attacks to the United States in 9/11, terrorism and the proliferation of weapons of mass destruction (WMD) further complicated the situation. Militarily, for the West, the first Gulf War in 1990, the war in Afghanistan launched in 2001 and the war in Iraq of 2003 revealed the fact that in this new international landscape, two different types of military operations emerged. The first emphasized the value of high-tech equipment and overwhelming firepower, both enabling the US and its allies to annihilate opposing regional forces. In this type of combat, the superiority of contemporary weapons systems was crucial for the military victory of the so-called 'clean war'. The second type of combat operations revealed the limits of Western superior technology in the post-war, stabilization phases unfolding from initial military victory. These time-consuming, laborious, manpower-intensive, and often bloody, operations in Iraq and in Afghanistan have exposed the difficult features of guerrilla warfare and counter-insurgency.

The experience of the past few years had as main implication that of forcing, to various degrees, developed nations to review their options for defence policy and international cooperation. These two new forms of war, so different from the conventional combat and traditional war-fighting ways of the Cold War, entailed that many armed forces had to develop more agile and flexible force structures and capabilities to meet new challenges. Similarly, governments and armed forces started exploring new ways to work together to address some of these issues, especially since many such crises involve forms of violence not aimed at one particular nation or group; rather, violence is aimed at trying to destroy the peace and prosperity of all human beings.

A similarly adaptable approach underpinned international military cooperation. Operations by multinational forces sought to take advantage of existing formats, but also had to develop new *ad hoc* mission-centred frameworks. These latter types of cooperation are not easy to conduct, and some form of an integrated mechanism among the participating military forces is necessary to enable – and guarantee – the safety and effectiveness of the operations. Recent experience in Iraq and Afghanistan suggests that effective integrated mechanisms depend on the alignment of military strategies, designating common military objectives, concrete burden sharing, standard operational procedures, evaluation of post action assessment, and extraction of lessons learned.

From a Japanese perspective, the advantages of an alliance in setting up such an integrated mechanism over a long period of time have some limitations too in meeting today's more fluid security environment. Alliances draw their main working mechanism upon the existence of a common agreed set of formal circumstances in which the alliance would be called into action, if the national interest of one related nation were not to concur with that of others in a particular instance, then it could be very difficult for the members of an alliance to reach an agreement and to conduct an operation. And the worst case could be the one where some member nations of an alliance do not participate in international operation driving the alliance towards its collapse. It is a common understanding of today's international community that in order to quickly respond to the dynamically changing situations by bringing all possible nations together, an alliance sometimes exposes its limitations.

A new form of international cooperation – which makes up for the limitations of an alliance system and is good enough to meet today's security requirements – is the coalition. Depending on the nature of a particular incident or crisis, each nation considers its participation weighing it against its national interest. In case of participation, the nation state still preserves the flexibility to determine details of its participating posture. No decision is taken under the binding legal framework of a treaty, or external pressure. Instead, the contributing nation state benefits from the advantage of having autonomy. In so doing, a coalition has the potential to invite more nations to participate in international activities and operations. As mentioned above, there is one fundamental factor that is necessary for a coalition to work in real operations. That is an integrated coordination mechanism to assure safe and effective multi-national military operations. And today, for example, an integrated coordination mechanism is constructed around US forces in the Middle East. The first Gulf War, the War in Afghanistan, the War in Iraq and the recently-started anti-piracy operations are typical examples of coalition operations and their associated coordination mechanisms.

Of course, the scope of this chapter is not to deny the value of an alliance. This would counter one of the core pillars of Japan's defence policy. The aim of the chapter is to underline that a wider approach to international military cooperation is essential to enhance readiness for any type of military operations at any time. In particular, as a former strategic planner, the worst scenario is one that includes a full scale head-on military collision among major powers and one that requires effective deterrence capabilities. An alliance is today the context within which these capabilities and force posture materialize. In case of a war breaking out, Japan should be strong enough to achieve final victory. What makes a goal of this kind true and viable for Japan is the country's alliance with the United States. For this reason, alliance and coalition operations are not mutually exclusive. They serve different purposes.

The Security of Japan: Coalitions Based on the Japan-US Alliance

In light of the changes of international cooperation, operating to meet challenges different from the type of crisis envisioned by the Japan-US Alliance, it is natural and practical for Japan to contribute through the form of coalitions. In fact, Japan has already participated in a support role during Operation Enduring Freedom (OEF) that operationally was in effect a coalition framework. Units of the JMSDF worked with and supported multinational naval forces for over eight years under this umbrella, and the frequent rotation of coalition navies interacting with Japanese forces really encapsulates the flexibility that many of them appreciated. Nonetheless, the fundamental operational basis of the JMSDF's overseas coalition operations remained the Japan-US Alliance. There are several reasons for this observation.

The first pertains to the historical precedents on the international operations of JMSDF. Since its foundation, the Japanese naval service has conducted operations and exercises with a variety of navies under the aegis of the USN. In particular, in its early days, the newly born, infant JMSDF had little experience in this field. As such, sets of information provided by the USN became a central part of its success. At the same time, the USN probably acted as an honest supervisor of the JMSDF, a remarkable fact given that this organization was the child of the IJN, the former formidable adversary. This form of cooperation and conduct based on a 'gentlemen's understanding' is still in effect today, and it has been a strong bonding agent between JMSDF and USN. This invisible and quiet professional bond has really been one of the most important factors that promotes and maintains the strong mutual trust and understanding between the two countries.

The second reason is related to the modalities of the JMSDF's introduction and understanding of the 'Standard Operational Procedures' (SOP) for coalition operations. The SOP for allied navies in the Cold War period had been developed within NATO and, whilst Japan was not a member of the alliance, its SOP were passed on to JMSDF by the USN since the 1950s. The SOP developed in this context served as the main reference for recent international naval operations under coalition framework. As such, for the JMSDF, it is still important to consult with the USN for various subjects on which JMSDF is unfamiliar. There are still many new subjects for JMSDF – both in terms of the conduct of operations and in the geographic areas of deployment. The JMSDF, by its upbringing and nature, has basically been an inherent part of a bilateral force in West-Pac for many years, and to retain its skills it maintains rich and complex training and operational experiences with the USN.

On the other hand, the Japanese experience of working with other navies still remains too limited, and the occasions have been too few. Thus, there are many areas the JMSDF needs to learn about, and the USN really becomes a 'good middleman' for that purpose. For example, there are still too many unknowns about operations in the Arabian Gulf and the Northern Indian Ocean, which are located so far away from Japanese waters. In addition to these, there is one factor that one should not forget: regional intelligence – a real Achilles' heel for the JMSDF outside its main operating theatre around the West-Pac waters.

On this matter, the JMSDF relies heavily on the USN, and solid intelligence is even more indispensable for coalition operations in distant waters. What really counts is that the USN has constantly offered access to the JMSDF – at any time and on any subject.

The third reason for the Japan-US Alliance to rest at the foundations of the JMSDF's increased participation to international coalition operations is the overall influence of the USN. The United States is a country that has undisputed influence on global security matters. For Japan, it is wise and reasonable to understand the implications and potential effects of this capability of the US, especially when the Japanese government has to pace its decision to participate in a coalition during an international security crisis. From the Japan-US Alliance point of view, areas around Japan and the West-Pac, as well as East Asia, have been considered to be those of core strategic and operational focus. Concurrently, Japan and the US have envisioned these areas as of primary concern for their security. This view of the alliance's focus is well understood and accepted in Japanese political and wider public spheres. By contrast, any area away from the West-Pac region may politically be considered to be removed from the area of responsibility of the Japan-US Alliance. The resulting view, according to some political analysts in Japan, is that the alliance should not work in distant areas from Japan.

However, this view is not correct. It is true that areas such as the Indian Ocean are technically outside of the West-Pac area of responsibility. Yet, this author would argue that the three reasons mentioned above clearly explain the importance of close coordination between Tokyo and Washington, even in case of a coalition in distant waters. The close consultation between the two nations and two navies really enhances mutual trust and understanding, and at the same time, it brings mutual benefits to both the JMSDF and the USN participating in coalition operations. This is why Japan has to coordinate closely with the USA, even in cases where JMSDF units were to be deployed to any security incident out of the West-Pac region.

What Cooperation Between Japan and the UK?

According to the new US Maritime Strategy, the naval forces of the United States have set homeland security as their first objective in the aftermath of 9/11. On the other hand, equally important goals remain the forces' nuclear deterrence, as well as their capability to engage successfully in a major conflict between nations. To achieve these strategic and military objectives, the build-up and maintenance of a military posture capable to project power far beyond national shores has been a central focus of American naval forces.

As such, the document connects homeland security to a specific means to secure it, pro-active action at the source of the security threat. The exercise of its 'influence' overseas is dependent on the country's 'global access' and in this context, the bases provided by NATO allies and Japan stand as indispensable tools to underpin US worldwide power projection capability. Among these overseas bases, those in Japan, located in geographical proximity to the East China Sea and South China Sea, and inbetween the US homeland and the Middle East, have been invaluable assets for decades. The bases in Japan, especially the naval bases and air stations whose functions stretch from operational to logistical support, include tremendous amounts of fuel and ammunition storage capacity. They are indispensable to give substance to the power projection capability of US Forces in the region and beyond. Without the bases in Japan, the United States and its armed forces might have had serious difficulties in exercising influence and securing access in various hot spots around the world, including the Middle East.

In this sense, power projection capabilities, supported by bases in Japan, constituted the backbone of US maritime strategy. This leads to the question of the relationship between the UK and Japan. As two major allied nations of the United States, how should these two countries proceed with future cooperation? What is important to consider here is the type of future cooperation with Washington and then the one between Tokyo and London. In developing the idea for this subject, one should take into consideration the global capabilities and posture of the United States, supported by its military strategy and capabilities geared towards the ability to exert power projection.

In addition, an idea for future cooperation should address the nature of international security and of crucial 'hot-spots' in crucial areas such as Southeast Asia, the Indian Ocean, and in the waters off the Horn of Africa. The reason why this area is selected as a model is that it is not only of current military and geopolitical interest, but also because the area has a future potential for our two nations to cooperate. There are three specific characteristics shared by both Japan and the UK. First, the maritime realm, occupying two-thirds of the planet's surface, provided both countries with a certain distance from their closest continental neighbours. This 'not too close, but not too far' nature provided and continues to provide common advantages and disadvantages for both Japan and the UK. Second, this maritime realm is home to significant interests for both nations, from trade and transportation of raw materials to fishing. So, for both Japan and the UK, the oceans possess a life-or-death type of importance if international instability were to affect trade and commerce. Last but not least, the oceans have a strategic significance for the scale of their national economic interests transcends the boundaries of their respective regions. This obliges them to keep a keen eye not only on the security and prosperity of their adjacent regions, but of global stability as a whole.

Taking these all three factors into consideration, the future course of action for Japan and the UK, and their navies – having different capabilities from those of the USN (whose focus is on Power Projection) – should be carefully considered. It is this author's belief that in most of the future security incidents at sea, both nations will participate in international efforts to control it. In this process, each nation may deploy its naval forces to the affected area based on its national policy, domestic situation and naval capability at the time. In the specific theatre of operations, the forces deployed will start cooperation with the USN, which is an alliance partner, and also between themselves.

A recent example is the JMSDF's OEF support force, which was deployed to the Arabian Gulf and Northern Indian Ocean, and cooperated with more than ten navies under close coordination with the USN's 5th Fleet and CENTCOM. The JMSDF support forces in the Indian Ocean thoroughly enjoyed opportunities to work with RN's warships, and they were really impressed by the seamanship shown by British sailors. Similar considerations are true for the anti-piracy operations off the coast of Somalia. Based on these two recent examples, a cooperative posture between the JMSDF and the RN units would represent natural development in future international operations. Of course, any cooperation like the humanitarian assistance and disaster relief operation conducted off of Northern Sumatra Island in December 2004 would also provide a great opportunity for both navies to cooperate. For Japan's and the UK's two nations and militaries, these types of cooperation in distant waters in military and non-military operations will surely be conducted mainly by naval forces.

Such an international cooperation would be designated as an 'expeditionary operation' or an 'expedition' by naval force. This notion is what draws the UK and Japan together. Whilst for the United States the key notion underpinning its global strategy is 'power projection', for

Japan and UK, two of its major allied nations, the crucial idea is 'expedition'. The political leaderships of both nations should understand this concept and should also create more opportunities for the JMSDF and the RN to take fully advantage of it.

The Way Ahead: The Future of Alliance and Coalition

In assessing the future, a strategic planner does have the luxury to become a fortune-teller looking into a crystal ball. Careful calculation, analysis, and a sound thought-process, based on accumulated facts and precedents, are the core ingredients of a successful formula. In the present security environment, the application of this formula suggests that full-scale state-on-state war between major powers is less probable. On the other hand, it seems more than plausible that future crises are likely to draw upon a non-conventional or asymmetric security incidents, a prominent scenario after the end of the Cold War. The numerous crises breaking out after the invasion of Kuwait in 1990 were such that many state actors in the West considered their degree of participation in a military coalition as the primary form of support for international crisis management and resolution. This flow of the tide towards coalitions emerged as a wisdom-of-life in the international community to meet the new and complex world of international affairs.

Within this trend, one additional factor to note relates to the fact that all the efforts to form coalitions and conduct operations under this framework were arranged under the United States, initiative and leadership. Of course, the United States, as a single superpower and the undisputed naval power in recent history, has significant stakes in any incident in the world, and also has heavy responsibilities to maintain global security and stability. These interests and responsibilities might be the driving factors for Washington to show initiative, and to take the lead in a coalition. In this respect, it is correct to acknowledge that US-led coalitions have been both practical and appropriate. For the same reasons, it is also practical to continue to envisage the formation of US-led coalitions in most envisioned future crises. This type of coalition has entailed, recently, and will continue to entail for both Japan and the UK the choice to favour 'coalition over alliance'.

What kind of other coalitions can strategic planners envisage for the future? The first case to take into account is one where the United States cannot, or will not, take the lead in forming a coalition. There are two possible scenarios in which this might happen. The first is one where the United States does not see any core national interests at stake in the international security crisis or incident at hand. In this case, Washington is likely to be reluctant to take the lead. The second scenario concerns a situation where another major power, or regional organization like the European Union, has stronger interests than the United States. In this situation too, the United States might be unwilling to take the lead to form and manage a coalition. Generally speaking, this second type of scenario would likely see the lack of shared interests, with state actors other than the United States expressing the inclination (or the necessity) to engage in the incident. In this kind of case, there are two potential state actors which might volunteer to fill the leadership gap, especially given their military potential: a reviving Russia, and an emerging China.

In this author's view, a coalition formed under the leadership of either country would not be desirable for Japan and UK, especially if such a coalition were to exclude the United States. In addition, from an operational operation point of view, such a coalition would also not be desirable because of a lack of well-established common SOP, an important advantage experienced by most navies participating in recent coalition operations. Such a lack of common established procedures could generate tactical confusion and some hazardous situations affecting operational safety. For this reason, taking these problems into consideration, the best and most practical solution – at least in the naval realm – is to form a coalition framework under the leadership of the RN.

The RN has been respected as a leading professional organization by all navies of the world. Even the Russian Navy or the Chinese PLA Navy cannot deny this fact. Furthermore, notwithstanding a decline in capabilities – well-documented in Professor Grove's chapter in this volume – the RN has been very effective in exercising its soft power. This in turn was essential to retain the ability to influence other navies. The possibility to draw the Russian Navy, the PLA Navy and other navies into a new coalition framework under British leadership, a new type of international cooperation would properly function. The United States would have little opposition to such an idea and, more importantly, such a solution would 'save the face' of both Russia and China from the risks of leading a complex naval coalition.

This proposition has no real precedent and, in light of the recent process drawing European countries closer in matters of defence and security, it might attract criticisms for being out of touch with current trends. Nonetheless, from the perspective of a Japanese strategic planner and

military professional, the European Union remains a rather elusive concept. Many European countries work together in a number of current
crises, and have worked with Japan in recent operations. Yet, here the
question is one of political and military leadership, and there is some
room to further examine the possibilities and potential of a British role
as *primus inter pares* within the European naval context. In particular,
what is important to all three navies, the JMSDF, the RN and the USN,
is that they should start to examine this proposal now, in other words,
well before a real incident actually occurs. There seems to be plenty of
time for the three navies to examine, to coordinate, and to develop their
common strategy to meet an undesirable international problem. The
UK and Japan, provided their close relationship with the United States,
should be prepared for this situation before it is too late. This proposal
could be designated as a strategy of 'coalition over kingdom'.

Conclusions

This chapter sought to articulate a Japanese view on international
cooperation at sea, taking into account the Japan-US Alliance as a core
element of the archipelago's national security. Against this background,
the chapter also briefly mentioned the historical ties and the current
status of the cooperation between the JMSDF and the RN. From a Japanese perspective, bilateral operations and exercises with the RN, small
in size and relatively few in frequency, have still provided precious
opportunities to improve the operational capability of the JMSDF.
Especially, taking full advantage of occasional visit of carrier battle
groups and nuclear attack submarines to Japan, the JMSDF has appreciated the unique operational experience gained through these exercises. They offered different lessons from those with the USN forces
stationed in Japan. In one pertinent example, an exercise with the RN's
battle group highlighted strong operational differences with the carrier groups centred on the USS *Kitty Hawk* or USS *George Washington*,
familiar to the JMSDF for many years. The experience of working with
the VSTOL carrier of the RN had a unique value to the JMSDF, especially since the service has very few chances to exercise with this type
of force in West-Pac waters.

Another symptomatic episode this author has vivid recollection
regards Japan's deployment of its minesweeping flotilla to Kuwaiti
waters in 1991. The RN's mine-countermeasures (MCM) units in the

theatre provided essential support to the JMSDF's MCM force. At that time, JMSDF's ships were in desperate need of a 'must have' type of magnetic field calibration prior to conducting actual clearance operations in a harmful minefield off Kuwait. The only naval unit capable of conducting this calibration was the RN's MCM support range in Dubai. Notwithstanding the absence of diplomatic arrangements to conduct mutual military support, the RN unit kindly granted the Japanese unexpected request and provided perfect support for the JMSDF's ships. Without this calibration support, the JMSDF's MCM operation off the coast of Kuwait could have ended in failure.

This episode seems small and less visible than other large scale operations. Yet, all JMSDF shipmates there, and in Japan at that time, really appreciated and applauded the true sense of common seamanship shown by sailors of the RN. Ashore, navy-to-navy senior staff talks between Ichigaya, the HQ of the JMoD, and Whitehall provide another good opportunity to exchange various strategic views and opinions between two geographically separated navies. These exchanges really made up for the 'handicap' of geographic separation. It is in light of the exchanges of these past years that this chapter sought to elaborate the notion of 'expedition' as a key concept for both the JMSDF and the RN. Similarly, the chapter articulated two new notions for Japan and the UK respectively to ponder as they seek to meet the security challenges of the contemporary world. 'Coalition over alliance' and 'coalition over kingdom' are the intellectual product of the present author's thirty-seven-year-long experience in the JMSDF. The hope is that, whilst the destination is impossible to fully predict, these ideas might offer guidance as to the future direction to all sailors of the JMSDF and the RN, and their political authorities in Tokyo and London.

CHAPTER TEN

CONCLUSION
MARITIME STRATEGY IN JAPAN AND THE UK:
THE 'ISLAND NATION' MODEL IN PERSPECTIVE

Alessio Patalano

> As a ruler, the east wind has a remarkable stability; as an invader
> of the high altitudes lying under the tumultuous sway of his great
> brother, the wind of the west, he is extremely difficult to dis-
> lodge, by the reason of his cold craftiness and profound duplicity.
>
> The narrow seas around these isles, where British admirals keep
> watch and ward upon the marches of the Atlantic Ocean, are
> subject to the turbulent sway of the west wind. (…) The north and
> south winds are but small princes in the dynasties that make peace
> and war upon the sea. (…) In the polity of winds, as amongst the
> tribes of the earth, the real struggle lies between east and west.[1]

The Maritime Strategy and National Security of Two Island Empires

The above passage, which appeared in the short story 'Rulers of East and West', was published by Conrad in April 1904, at a time when the 'east wind' had fallen upon 'his great brother'. In February, a surprise night attack by a squadron of Japanese destroyers against Russian naval forces based in Port Arthur had prompted the first naval engagement of what came to be known shortly after as the Russo-Japanese War. This conflict became the first modern struggle between 'east and west'. In retrospect, a contemporary reader is left wondering as to whether Conrad's creativity had gone as far as to consider the possibility that the east wind would eventually prevail over the wind of the west. What is more certain though, is that his metaphorical reference to British admirals as keeping 'watch and ward upon the marches of Atlantic Ocean', perceptively captured one essential aspect of British national

---

[1] Joseph Conrad, *The Mirror of the Sea & a Personal Record* (Ware, Hertfordshire: Wordsworth Editions, 2008), 95.

security priorities at the turn of the century. An aspect that is directly relevant to what John Ferris and Tohmatsu Haruo argued in the opening chapters of this book.

As Ferris pointed out, at the beginning of the twentieth century, British economic wealth and political independence rested on two main pillars. The first was the British imperial system: a global economic network made of colonial trading outposts connected by oceanic shipping lanes and defended by a powerful naval force. Some authors regarded this as the 'imperialism of free trade', one that was driven by a quest to access markets and resources on a global scale and less prone to territorial occupation and political control.[2] Global trade and finance were the lifeblood of British economic power and in turn, the empire connected the shipping lanes that delivered the country's wealth. Such a dependence upon the sea meant that Britain had to maintain superior naval power, either directly by means of a powerful fleet with global reach or indirectly by means of regional partners, to guarantee its economic survival.

Indeed, the second pillar was directly connected to this question of 'how' naval supremacy was to be maintained. In the words of British imperial historian Gerald Graham, naval supremacy required, first of all, an unquestioned control of the 'narrow seas around these isles'. This was essential to enable Britain 'to watch, control, and often demolish the main naval forces' of its main European competitors, especially France, Russia and, later on, Germany.[3] This mission was of paramount importance and it conditioned the country's approach to the question of imperial defence. The Anglo-Japanese Alliance that was signed in 1902 was the result of the need to meet the requirements of imperial security. At a time when the margins of British technological and numerical superiority were under pressure, the need to concentrate core naval capabilities around the channel to prevent invasion and keep European competitors in check made the presence of a regional ally to defend the remote corners of the empire an appealing move. In this context, the alliance was not the result of a natural affinity between two 'Island Empires'; it was part of a maritime strategy that served the interests of a world-wide empire.

---

[2] The concept of 'imperialism of free trade' mentioned here is explored in Ronald E. Robinson & John A. Gallagher with Alice Denny, *Africa and the Victorians: The Official Mind of Imperialism* (2nd Edition, London: Macmillan, 1982).
[3] Gerald S. Graham, *The Politics of Naval Supremacy. Studies in British Maritime Ascendancy* (2nd Edition, Cambridge: Cambridge University Press, 2008), 23.

Professor Tohmatsu further expanded on this subject explaining how the pursuit of a bilateral relationship served well not only British strategic requirements, but also Japanese regional ambitions and security concerns. The alliance system gave Japan privileged access to the world's most exclusive international political circles. It contributed to raise its international profile from regional player (a status gained after the success in the First Sino-Japanese War) to major power (especially in light of victory over Russia). On the subject of Russia, the alliance directly benefited Japan by helping the country to address vital security concerns, Russian influence in Manchuria and Korea and, later on, German economic and military encroachment in China and the Pacific area. Japanese naval capabilities were an essential asset that the country brought to the partnership with Britain. As Tohmatsu examined in his chapter, this became particularly evident during the First World War. The Japanese ability to operate across the Asia-Pacific region underpinned British requests for different kinds of military assistance, ranging from the contribution of ground forces to defend parts of the empire like India, to naval assets to protect the sea lanes towards Australia, and in the Mediterranean. In all, sea power was a core constituent of the Anglo-Japanese Alliance, serving well the requirements of two different maritime strategies and national security policies.

These differences became evident in the 1920s. The threats that had originally favoured an Anglo-Japanese strategic alignment had been removed from the list of the two countries top national security concerns. This shift, combined with the emergence of Japan as the prime regional military power in East Asia, contributed to a change in mutual perceptions. The substitution of a bilateral system with a multilateral treaty marked the end of the original strategic alignment and the emergence of this change in perception. From that point onwards, Japanese sea power was the main parameter underscoring the country's transformation from a valuable regional ally to a potential competitor. Throughout the 1930s, this remained an unresolved dilemma, and British understanding of Japanese military power led authorities in London to deem their counterparts in Tokyo as lacking the capabilities to mount a vital threat to imperial possessions in East Asia. As Douglas Ford pointed out, British leadership was faced with the difficult choice to assess Japanese war-making capabilities against increasing financial pressure at home and growing political tensions in Europe. In this context, the strengthening of the Singapore naval base as the pivot of imperial defence was a solution that played to the strengths of the British imperial system. In the second half of the 1930s, the pace of Japanese preparations stretched beyond the ability of British intelligence to fully appreciate them. In this respect, one important point raised in his chapter is that the fall of Singapore was not to be ascribed to an overall unsound strategy. Rather, it depended on a limited preparation based on a degree of under-estimation of Japanese capabilities and intentions.

This is a particularly relevant point that draws the discussion in the first part of the book back to John Ferris' conclusions about the different relationship existing between maritime strategy and national security in Japan and Britain before the Second World War. By the end of the nineteenth century, East Asia represented a 'maritime system'.[4] In geographic terms, the sea constituted the region's main connecting fabric with shipping routes providing highways for trade, transport, and the projection of power across its most remote corners. As Ferris pointed out, the high degree of 'connectivity' of this regional system served different strategic purposes in Japan and Britain. British national security put a premium on the inherent role that East Asian waters had in supporting the economic activities at the foundations of the imperial enterprise. The projection of power was pre-eminently a function of the need to safeguard global shipping routes. For Japan, on the other hand, regional waters represented a means that made Japan accessible to attempts of foreign expansion first and a means for territorial expansion later. In the Japanese case, the emphasis was on the military use of the sea. Provided East Asian geography, sea power was strategically important to both Japan and Britain. Only in the case of the latter, the need to build-up and maintain it was inscribed into a strategy that put military capabilities at the service of a wider national economic entrepreneurship.

This is not to say that, by the time the war in the Pacific took the form of a protracted effort, Japan's own economic survival did not increasingly depend upon the sea. The raw materials available in Southeast Asia became essential to propel the Japanese war machine. In 1942, Japanese military forces managed to recover 70% of the East Indies pre-war production rate of 180,000 barrels per day and by the end of the following year, Japanese oil imports in absolute terms had increased from 29,000 barrels daily to 40,000. In 1943, the sea routes of Southeast Asia

---

[4] For an explanation of the different nature of continental and maritime systems, cf. Jack S. Levy and William R. Thompson, 'Balancing on Land and at Sea: Do States Ally against the Leading Global Power?', *International Security*, Vol. 35, 2010:1, 16-19.

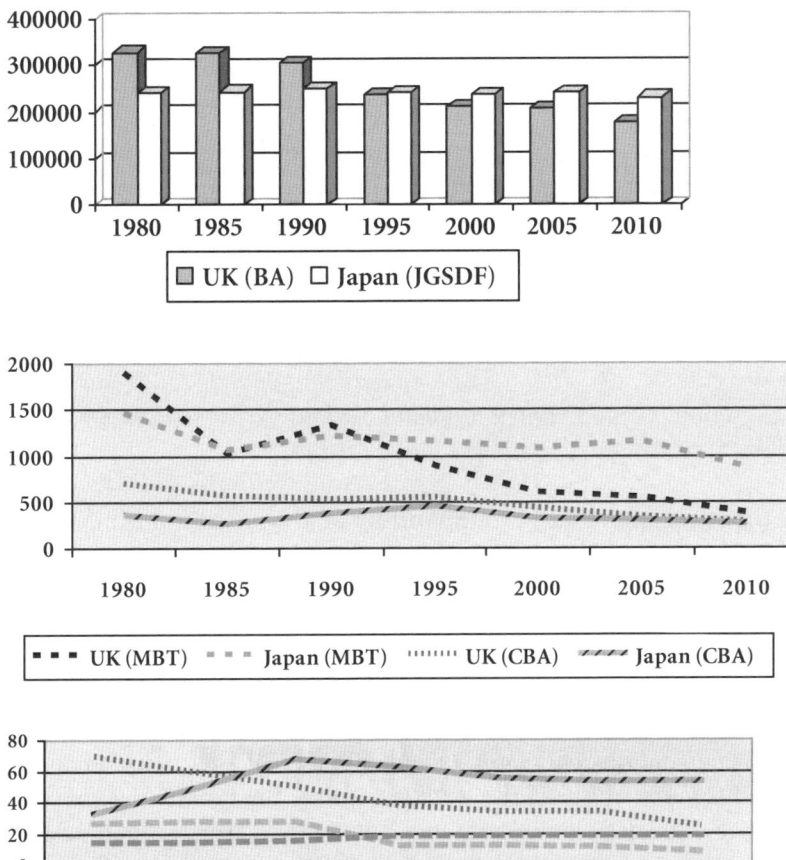

Figure 10.1. The transformation of Japanese and British military power, 1980-2010
Source: International Institute for Strategic Studies, The Military Balance (Various Years, 1985-2011).

represented a central artery to fuel Japanese naval assets.[5] The region produced 136,000 barrels of oil daily, 96,000 consumed by fleet assets regionally and the rest exported to the Japanese homeland. In 1944

---

[5] In the period 1942-1944, the IJN fought three major battles, Midway, the Philippine Sea/Marianas and Leyte Gulf. These three alone consumed over 1 million tons of oil. Mark P. Parillo, *The Japanese Merchant Marine in World War II* (Annapolis, MD: Naval Institute Press, 1993), 28.

and 1945, the Allied naval war effort was progressively redirected to harass Japanese lines of communication. This change of strategy had a devastating effect. Japanese oil imports fell from a 1943 peak of 740,000 tons to a meagre 178,000 tons in the third quarter of 1944. A mere 9% of oil shipments from Southeast Asia reached their final destination in Japan in 1945.[6] The water masses of the Asia Pacific had ceased to be a Japanese lake but were turned into a graveyard of critical raw materials. In Ferris' words, the defeat Japan suffered in the Second World War was symptomatic of a failure to 'master or to reconfigure the Asia-Pacific system'.

From 'Island Empires' to 'Island Nations': Japan and the UK in the Post-War Era

In the aftermath of the Second World War, the sea continued to have a significant role in the way Japanese and British defence policy-makers debated strategy and national security. In a way not too dissimilar from the pre-war years; in the Cold War too political and military elites in Tokyo and London had to make difficult choices to harmonize insular geography, peacetime economies, and reduced military power.

In Japan, General Yamaguchi's chapter presented an evolution of threat perceptions and military capabilities that was symptomatic of a re-discovery of 'insularity' as a way to set the boundaries for 'self-defence'. Especially in the early stages of the Cold War, the archipelago's geographic separation from the neighbouring continent contributed to physically define where the new notion of national defence applied. A new normative framework prohibiting the use of war as a tool of foreign policy, the signing of a bilateral alliance with the United States to guarantee a nuclear umbrella over the archipelago, and the prioritization of economic security further reinforced the use of geography to limit Japan's efforts in defence affairs. Japanese military power was pre-eminently tasked to repel an invasion of its main islands.

By the end of the 1970s, this approach progressively expanded under the combined action of increasing concerns for Japanese shipping and energy security and pressure from the United States. 'Insularity' came to mean more than just physical separation and a natural wall for the Japanese forces to defend. As the Japanese economy grew stronger and

---

[6] Ibid., 247, 215.

more dependent on access to foreign markets and energy resources, insularity became synonymous with vulnerability and dependence upon the sea for economic survival. This new meaning led Japanese politicians to commit to a defence policy beyond national shores in an attempt to secure the main sea lanes approaching the archipelago up to a 1,000nm. This policy shift *de facto* consolidated the role of ground and air forces to homeland defence in case of an increasingly unlikely invasion of the northwest part of the archipelago. Concurrently, it set maritime forces on the trajectory to emerge as the primary tool to protect wide-reaching national economic interests. The adoption in 1976 of the 'Basic Defence Force' concept crystallized this vision into a doctrinal grammar that shaped military capabilities for the reminder of the Cold War and eventually informed the debate of the subsequent two decades.

Just as the policy actions were fully implemented, changes in the international system introduced new variables for Japanese defence planners to consider. As General Yamaguchi and Professor Aoi remarked, the key questions of the post-Cold War period became: did Japan possess the right military balance and normative provisions to take part in international military operations? To what extent was it in Japanese security interests to intervene in missions aimed at maintaining international stability? Professor Aoi suggested that the answer to the first question took time, and the time required depended on the JGSDF's initially limited operational experience overseas. For this reason, the operations in Iraq had a major impact on the JGSDF. By the end of the two year-long deployment, it was clear to Japanese military authorities that there was a 'significant gap between the country's material capabilities and the "software" of its operations'. This was particularly true for the ground forces since they had not been designed for operations outside the Japanese main islands. Shortcomings were identified in command structure, organization, and logistics. As Professor Aoi noted, the recent establishment of the CRF commanding a regiment of 1,300 men ready to deploy ahead of larger contingents, represented a consistent step to enhance Japanese military capabilities for international operations.

Both General Yamaguchi and Professor Aoi emphasized how the upgrading of international peace cooperation to the status of core mission for the JGSDF happened in a context where conventional sources of concern in the region persisted. The question of 'direct attacks to the archipelago', as General Yamaguchi put it, required the

Japanese defence establishment to review capabilities and normative frameworks to better address the increased number of incursions into Japanese territorial waters and invasion of national air space, by conventional forces (submarines, spy boats, etc.) and missiles. Conventional threats remained a central factor in the definition of missions and capabilities of the Japanese military. This is clearly explained in the NDPG 2004 where 'new threats and diverse situations' included responses to ballistic missile attacks, to guerrillas and special operation attacks, to invasions of off-shore islands, to surveillance of maritime and air space, to major disasters.[7] In the NDPG 2010, the result of two decades of debate over how to strike the appropriate balance between expeditionary capabilities and the defence of the archipelago was summarized in the doctrinal shift from the 'Basic Defence Force' to the 'Dynamic Defence Force' concepts. The former placed priority on a passive approach to the use of military power, 'ensuring deterrence through existence *per se*.' The latter emphasized performance, operational flexibility and enhanced systems integration, 'placing importance on dynamic deterrence'.[8]

Guibourg Delamotte further pointed out that the progressive recalibration of the JSDF was conducted alongside a political struggle over the interpretation of what the Constitution would allow Japan to do in the international arena. In this context, it is perhaps not a coincidence that in regional waters as well as overseas, maritime operations acted as a spearhead for change, stretching the boundaries of constitutional limitations, and as a 'laboratory' to test new 'special' normative provisions. From the deployment of the first flotilla of minesweepers in the aftermath of the First Gulf War, to the operations in the Indian Ocean and in the Gulf of Aden, to the incursions of foreign vessels in Japanese waters, the JMSDF led the way for change, raising the thresholds for the political acceptance of new missions. The relative isolation and self-containing nature of operations at sea enabled Japanese political elites to debate the normative limits of Japanese operations without ever requiring a comprehensive reconsideration of sensitive notions such as 'collective self-defence'. From a normative perspective, the sea empowered Japanese elites with the flexibility to shape the Japanese contribution to international stability, and decide the country's degree of political and military engagement.

---

[7] JMoD, *National Defence Programme Guidelines for FY 2005 and Beyond*, 4-5.
[8] JMoD, *National Defence Programme Guidelines for FY 2011 and Beyond*, 7.

In post-war UK, defence policy followed a different path. Victory in the Second World War left the country in a difficult situation. International responsibilities and security interests unfolding from the country's imperial status and as a key member of the UN National Security Council and NATO were not matched by the economic power necessary to pay for a military structure with global reach. As Eric Grove noted in his examination of the evolution of the RN, the possibility to retain expeditionary forces and a degree of power projection capabilities was essential for the UK to consistently 'punch above its weight'. British interventions in the wars in Korea, the Falklands, and the Gulf, and in more recent times, operations in Yugoslavia, and Sierra Leone, proved this point. On the other hand, Professor Grove's account exposed the too-often assumed notion that for an 'island nation' a certain natural 'awareness' to the advantages of a maritime strategy is self-assured. Since the early stages of the Cold War, the success stories of the post-war era were the result of a national military balance in which the navy displayed the skills to manage and rejuvenate capabilities left over from the fleet of the imperial and wartime experiences in the face of defence policies with a limited 'maritime' outlook. The history of post-war British naval policy is in fact characterized by the constant need for the navy to fight fierce procurement battles to guarantee the survival of its assets, with its role in security and foreign policy constantly questioned. Looking at the comparative tables reproducing the evolution of British and Japanese military power, one cannot fail to be struck by the substantial and constant reductions that have occurred over the past thirty years.

Bringing the narrative into the challenges of the UK's more recent military commitments, Commodore Jermy added to Eric Grove's considerations, and explained the impact of the past ten years of operations on British defence strategy. Onerous deployments in Iraq and Afghanistan, with constant public visibility and political attention to British personnel operating in the battlefield of the Iraqi desert and of the Afghan mountains, drove the defence debate away from matters of grand strategy. Discussions over the conduct of operations and the procurement of equipment hindered the debate over the future balance of British military power. In this context, Commodore Jermy's proposal of an off-shore strategy strikes the reader as an attempt to rebalance the national security equations by advocating to take advantage of a naval/air-centred military posture to deter potential threats far from the British Isles, exert international influence, whilst retaining in the meantime the control of the degree of involvement in a particular crisis.[9] This proposal seemed to suit the present British security environment, where the lack of conventional threats in immediate European continental neighbourhood reduced the need for a large army. The increase of tensions in areas of key economic interest to the UK, by contrast, would put an island nation like the UK in the ideal position to opt for the return to expeditionary forces tailored to project power and conduct specific tasks on the ground such as SSR.

In both chapters on post-war UK, the alliance with the United States is presented as a significant factor in British security policy. In terms of capabilities and combat systems, a close strategic partnership with the United States enabled the British armed forces to access advanced technology, the most notable example being the development of the RN's submarine-based independent nuclear deterrent. In the realm of doctrine and tactics, exchanges with the United States within and outside the NATO context, exposed British forces and assets to the latest developments on carrier aviation and ASW operations, to offer just two examples. Politically, a security alignment with the United States offered the possibility to the UK to continue exercise influence at the international level notwithstanding the progressive reductions of its military capabilities and especially after the British withdrawal 'from East of Suez' in the mid-1960s. Here, Colin Gray's idea mentioned in the introduction about British strategy as one based on the combination of national capabilities and the alliance with a strong naval power is consistent with the security policy of the post-war period. This does not mean, as Commodore Jermy argued, that the advantages brought about by the relationship came without costs. The adventurism of the Labour governments led by Tony Blair and the support given to the American-led War on Terror put an enormous stress on British military capabilities setting in motion a policy review that was concluded in 2010 and that considerably reduced British power projection capabilities in the short and medium terms. In this respect, it seems almost ironic that the crucial role played by British power projection in supporting the country's Cold War ambitions and responsibilities firmly alongside the United States, is now being so affected by alliance commitments.

---

[9] The latest policy document produced by the RN articulates the navy's future strategic context along similar lines. Cf. Royal Navy, *The Royal Navy Today, Tomorrow and Towards 2025* (Royal Navy: Fleet Graphics Centre, 2011), 2-3.

In all, what then of Japan and the UK and their transformation from 'Island Empires' to 'Island Nations'? Until the mid-1970s, Japan and the UK followed two opposite trajectories in the way geography informed national security. In Japan, the defeat brought about demilitarization and the need to focus on the reconstruction of a shattered economy and the reintegration into the international system. The alliance with the United States was an important means to achieve such aims, a fact well-explained by Admiral Kōda in his chapter. In security terms, Japanese political authorities could count on the country's insular geography to limit military expenditures to the procurement of capabilities to defend the home islands. In the UK, victory in war came at equally considerable costs. This affected the peacetime reconfiguration of national military power, conducted with the goal to reduce defence spending whilst seeking to maintain an international profile consistent with British reputation and influence. The flexibility and reach of British military power helped this transition from a global empire to major player in the international arena or, as Richard Hill had it in the 1980s, to the status of 'medium power'.[10] The use of the oceans as a staging platform from where to exert power and influence, enhanced by a network of suitable bases overseas, were central to British success in war and peace during this period. For the UK too, however, the alliance with the United States – as well as the British role in NATO – compensated for the limits of national capabilities. Insularity informed the security policies underpinning Japanese post-war recovery and British post-imperial transition.

By the mid-1980s, the gap between Japanese and British military power had been substantially reduced. Especially in the maritime realm, the two countries featured fleets increasingly similar from a numerical point of view. Functionally, the Japanese fielded a force pre-eminently conceived to conduct ASW operations in the East Asian region, in the Sea of Japan and along the main sea routes approaching the country from southeast and southwest. The UK possessed a more balanced fleet, one that was suitable for ASW and expeditionary operations away from national shores, as the Falklands War proved. Overall, Japanese security policy was becoming more maritime in its outlook, the UK's less so.

A little more than a decade and a half later, the original gap is on the brink of disappearing. Japan is reconfiguring its military capabilities

---

[10] Richard J. Hill, *Maritime Strategy for Medium Powers* (Annapolis, MD: Naval Institute Press, 1986), 25. Admiral Hill updated his original ideas in R. J. Hill, *Medium Power Strategy Revisited* (Working Paper N.3, Canberra: Royal Australian Navy Sea Power Centre, 2000).

Table 10.1. Japanese and British military capabilities

	United Kingdom	Japan
Total Personnel	178,470	247,746
(Reserves)	82,274	56,379
Army	102,600	151,641
MBT	325	850
RECCE	738	100
Helicopters	66 (ATK) 450 (ISR) 232 (MRH)	185 (ATK) 118 (ISR) 236 (TPT)
Navy	35,480	45,518
SSN – SSBN	4 – 7	—
SSK	—	18
CVS/CVH	1	1
CGHM	—	2
DD	7	30
FFG	17	16
MCM	16	37
Amphibious	3 (4 RFA – Bay Class)	3 (2 Small LST)
Aircraft – Helicopters	12 – 132	95 – 130
Air Force	40,390	47,123
Aircraft	334	374
Helicopters	175	53

Source: International Institute for Strategic Studies, *The Military Balance 2011* (Abingdon, OX: Routledge for IISS, 2011).

to make them more effective, flexible and deployable. Today, deployments overseas like the anti-piracy mission in the Gulf of Aden are conducted whilst enhancing a credible conventional presence in the Sea of Japan and in the East China Sea. A maritime strategy combining ASW, ISR and amphibious capabilities to defend regional sea lanes, national straits and off-shore islands is essential to the Japanese need to balance national and international commitments.[11] The major exercise

---

[11] Patalano, 'Japan's Maritime Past, Presence and Future', 104-105.

organized by the JSDF in the Amami group of islands (Okinawa prefecture), started on 10 November 2011 and involving about 35,000 SDF members, 1,300 vehicles, six warships and 180 airplanes, is an example of Japan's efforts to implement such a strategy.[12] The UK, on the other hand, is finding it increasingly difficult to keep the past military balance. Indeed, as a British academic remarked as the country was involved in the air and maritime operations in Libya, the coalition government 'has ground to dust most of the key enablers for the sovereign projection of British power'.[13] The 'enablers' he referred to in the article were the aircraft carrier *Ark Royal* and its air wing of *Harriers* jet fighters. Militarily, Japan and the UK started their post-war journey in very different places but, at the end of the first decade of the twenty-first century, they seem to have ended in very similar ones, with Japan quietly taking a clear lead in some capability areas.

For both Japan and the UK, sea power never ceased to matter. Either as a defensive wall or as a means to engage with the international environment, the question of the role of an insular geography featured prominently in national security. Sea power underpinned debates on national strategy addressing the balance between land and naval/air power. In some cases, political choices favoured a stronger land-centred balance to defend the Japanese archipelago and, in the case of the UK, the European continent. In more recent times, naval forces received greater attention as national security was better served by the ability to operate farther from national shore, projecting power in the form of leaner, more flexible expeditionary forces. Sea power mattered in the debates over the extent to which national military capabilities had to be complemented by bilateral and multilateral alliances. It mattered because shifts in the power structure of world politics did not make the dependence of Japanese and British economies on trade and maritime shipping irrelevant. The harmonization of geography and military power to pursue economic benefits meant that operating from the sea and at sea remained for Japan and the UK not a question of 'if', but a question of 'how'. Last but not least, sea power mattered in the management of the alliance with the United States, not only because it

---

[12] Editorial Office, 'SDF starts unprecedented exercise in Kyushu, Okinawa', *Asahi Shimbun* (Asahi Japan Watch), 11 November 2011, http://ajw.asahi.com/article/behind_news/politics/AJ2011111117195, accessed on 12 November 2011.

[13] Gwyn Prins, 'If Britain Wants a Say in Global Affairs, It Needs Muscle', *The Telegraph*, 3 August 2011, http://www.telegraph.co.uk/news/uknews/defence/8677550/If-Britain-wants-a-say-in-global-affairs-it-needs-muscle.html, accessed on 12 November 2011.

made the two countries valuable strategic partners, but also because it enabled Japanese and British governments to retain a degree of political independence.

Looking Ahead: From 'Island Nation' to 'Island Nation Partners'?

In the latest round of the debate on whether the post-Cold War changes of Japanese defence policy are a sign of the country's achieved status as the 'Britain of the Far East' or, as other authors have it, as a 'global ordinary power' based on the UK model, one Japanese author expressed a negative opinion. He argued that Japan is not and is unlikely to be like the UK, 'a state that does not hesitate to conduct combat operations jointly with the United States or even independently'.[14] This book showed that this type of notion is based on a partial interpretation of the analogy and it argued a more articulated comparative analysis between Japan and the UK, one based on the 'Island Nation' model.

In the realm of security, being an 'island nation' leads Japan and the UK to share the following characteristics:

INSULAR GEOGRAPHY. An insular geography is both a strategic asset and a critical vulnerability. It is a strategic asset in that Japan and the UK benefited (and continue to benefit) from the physical separation from their respective continental neighbours. This enabled their political and military elites to choose the degree of engagement in continental affairs necessary to guarantee independence and territorial integrity. The nature of the engagement in continental affairs informed the configuration and reconfiguration of military power and capabilities, especially in regard to the size and shape of ground forces. Insularity represented also a critical source of structural weakness because both Japan and the UK depended (and continue to depend) on the use of the sea to develop and maintain their industrial capability. Japanese and British economic activities, from the export of manufactured goods to the import of food supplies and raw materials, to the exploitation of natural resources, rest on access to the sea. Disruptions in the maritime transportation system are, for an island nation, a fundamental strategic vulnerability.

---

[14] Yasuhiro Izumikawa, 'Explaining Japanese Antimilitarism: Normative and Realist Constraints on Japan's Security Policy', *International Security*, Vol. 35, 2010:2, 158.

GLOBAL INTERESTS. The security interests of an island nation tend to be connected to global affairs and the uninterrupted access to foreign markets and resources. This was true for the British Empire as it is for post-war Japan. The 2010 NDPG reminded its readership that this is a core premise upon which defence policy is formulated. 'Japan (…) is a trading nation which heavily depends on imports for the supply of food and resources and on foreign markets. Thus, securing maritime security and international order is essential for the country's prosperity'.[15] In this context it is worth noting that in the 2010 SDSR, reference is clearly made to the 'global' nature of British interests and ambitions. In the document, however, the meaning of such a reference is never fully explained other than in terms of a 'proud history of standing up for the values we (the British people) believe in'.[16]

MARITIME-INFORMED DEFENCE POSTURE. A maritime strategy constantly informs the defence posture of an island nation, shaping the character, composition and reach of national military power. A maritime-informed defence posture is not equal to saying that island nations consistently opt for a military balance centred on a strong navy. On the contrary, structural considerations pertaining to the character of the threats emerging from the continental neighbourhood and the state of the national economy contribute to determine how the balance between ground and maritime forces was to be struck. This last factor is particularly relevant since navies are technology-intensive organizations and their assets tend to be expensive, requiring continued financial investment for maintenance, upgrades and eventually, replacement. In the case of Japan this point leads to an additional consideration. The analysis in this book of the impact of constitutional limitations on defence posture suggested that these have affected the nature of military operations in relation to the use of force. They seem to have less of an impact, however, on matters of grand strategy. Island nations are not interested in territorial gains as much as they are in international stability and safe transportation at sea. Their economic lifelines are at sea, not on land. Japanese defence policy is today underscored by a maritime strategy that seeks to achieve the two primary security goals of an island nation: the defence of territorial integrity and the safeguard of economic interests.

---

[15] JMoD, *National Defense Programme Guidelines for FY 2011 and Beyond*, 4.
[16] HM Government, *Securing Britain in an Age of Uncertainty: The Strategic Defence and Security Review* (Cm 7948, London: HMSO, 2010), 3.

ASYMMETRIC MILITARY STRUCTURES. Island nations like Japan and the UK tend to possess relatively compact military structures – usually with an asymmetric structure focusing on small armies and larger air/naval forces. The advantages in terms of limited requirements for territorial defence due to the insular character of national geography are offset by the necessity to address questions of security of the maritime realm. The extent to which an island nation can build naval and air forces capable to engage with the defence of national interests at sea will depend on its economic power. As a result of the asymmetric nature of the armed forces' structure and of the costs of naval and air capabilities, island nations tend to underfund their military power in peacetime. Limited funding is counterbalanced by constant attempts to maximize the projection of power by using the sea as a vast manoeuvring space. For this reasons, island nations tend to be capable of 'punching above their weight' – although, as in the case of contemporary Japan, they might decide not to. In this context, it is interesting to underline that both in the case of Japan and the UK, major changes to the asymmetric vocation of their military structures to favour large land commitments, substantially weakened their defence capabilities. The examples of Britain in the two world wars, Japan's war in China, and the UK's recent commitments in Iraq and Afghanistan are all cases in point.

PARTNERSHIP-ORIENTED SECURITY POLICY. Island nations tend to rely on close strategic and military partnerships to complement the limits of national military power. Especially in the post-war period, the United States represented for both countries an essential military partner, offering access to know-how, hardware, tactics and doctrines. In the case of the UK, the existence of a military alliance such as NATO further added to national power and influence. For both Japan and the UK, the benefits of a strong alignment with the United States outweighed the costs of its management, though in recent times, these have been considerably higher. The wars in Iraq and Afghanistan and the relocation of American bases in Okinawa stand out as examples of the higher costs faced by the UK and Japan respectively. The requirements for Japanese and British security policy to maintain a close military relationship with the United States is consistent with the limits and vulnerabilities of an island nation and as a result, is likely to remain unaltered. Neither country possesses the economic foundations to pursue independent military capabilities capable to ensure the safeguard of their national interests. As island nations, both countries are likely to invest more to expand the portfolio of military partnerships, seeking issue-based forms of cooperation, especially in areas of key strategic interest to reduce the weight of the management of the alliance.

This last consideration takes the concluding remarks of this book to address one final question. What about Anglo-Japanese military relations? As the narratives in some of the chapters in the second and the third parts of this book indirectly suggested, neither country scored high in the post-war defence policy priorities of the other. In the 1960s, security cooperation in Southeast Asia to be developed together with members of the British Commonwealth, and British defence industry opportunities in Japan briefly entered the respective national agendas.[17] These remained somewhat peripheral initiatives in Japanese and British national security policies. In retrospect, provided today's similarity between the two countries and the years of regular but limited contact at the senior military and defence officials levels, a closer scholarly investigation of the post-war evolution of the military exchanges would be of great benefit to understand the potential challenges to future cooperation.

This is particularly true since a new trend seems to be emerging. Over the past decade and a half, in military affairs, senior staff talk meetings have been reinforced by important official visits, most notably the participation of the JMSDF's training squadron in the celebrations for 'Trafalgar 200', and occasional joint training activities. In particular, as this book goes to press, Japan, the UK and the United States have just completed a mine warfare trilateral exercise in Bahrain. This is a significant activity as it is the first of its kind to take place among the three countries in the Middle East, in an area not too far from where they have all been involved in the protection of international trade through the anti-piracy mission. The Japanese deployment of two large minesweepers for a total of approximately 180 personnel is a clear sign of the importance assigned by the country to the exercise and the prospects of close cooperation in the area with other countries in addition to the United States. Industrial relations too witnessed new developments during this period. As of 2003, the JMSDF procured a major combat system, the Augusta Westland EH101 (formerly AW101) helicopter, in

---

[17] Christopher Braddick, 'Distant Friends: Britain and Japan since 1958 – the Age of Globalization', 263-275, in Nish and Kibata, *The History of Anglo-Japanese Relations, The Political-Diplomatic Dimension*; also C. Braddick, 'Britain, the Commonwealth, and the Japanese Post-war Revival, 1945-1970', *The Round Table*, Vol. 99, 2010:409, 371-389.

both MCM and transport versions to replace its ageing fleet of MH 53E and Sikorsky S-61. In the realm of defence industry, another major combat system, the Eurofighter Typhoon aircraft, was one of the top three candidates in the recent bid for replacing Japan's fleet of F-4 Phantoms.

Whilst none of these developments represent an ultimate guarantee of a political commitment to enhance defence cooperation, they are all symptomatic of shifts in international security that should alert Japanese and British political and military elites to the importance of considering closer ties. Admiral Kōda's concluding chapter on the state and prospects of Anglo-Japanese military relations offered crucial insights on a Japanese perspective on this subject. As island nations, Japan and the UK share an interest in the stability of the international trade system, and in the prevention and/or containment of threats to it. As Admiral Kōda suggested, neither country wishes or has the military power to take up the role of 'global policeman' in a fashion similar to the United States. Both countries, on the other hand, possess military structures that can allow their political leaders to intervene in critical hot spots and contribute to international efforts in areas of strategic importance, including the Horn of Africa, the Indian Ocean, and Southeast Asia.

In the face of limited national military capabilities, joining forces on the basis of common strategic needs is a path consistent with the strategy of island nations. As this book has shown, Japan and the UK have more strategic commonalities today than ever before, and these should be regarded as a crucial incentive for collaborations in specific areas of the world, on specific type of operations, on the development and procurement of combat systems, on the adoption and implementation of practices and procedures. The Japanese increased presence in the Horn of Africa, with the opening of their first overseas base in Djibouti is a sign of this island nation's approach to international security. The UK's presence in the Persian Gulf and in the Indian Ocean is similarly representative of the country's 'maritime' approach to strategy. A similar longstanding military partnership with the United States would facilitate the implementation of closer operational ties whilst, at the political level, support for British operational leadership in joint operations would be welcomed in Washington as much as in London and Tokyo. In a context where the United States faces challenges to its global supremacy, the enhancement of strategic ties between two of its closest partners would fill potential leadership and political gaps. Concurrently, the shared principles upon which national security policies are formulated in Japan and the UK would make it easier to identify projects for industrial cooperation which would in turn strengthen national military capabilities and increase their credibility as international players and reliable allies.

In all, Japan and the UK both stand at a crossroads. Through different historical journeys they are today island nations facing the difficult task to contribute to international stability to guarantee safeguarding their well-being. As island nations, they both have characteristics that make them similar to each other and different to other state actors. As island nations, they have limited national military resources but can use them to exert considerable influence at the international level. As island nations, it is by building on the maritime approach underpinning national security that they can find a way to develop stronger ties and remain abreast with the change of tides of international security affairs.

INDEX